# THE NATURE AND PROSPECT OF BIOETHICS

# THE NATURE AND PROSPECT OF BIOETHICS

*Interdisciplinary Perspectives*

Edited by

## Franklin G. Miller
*National Institutes for Health, Bethesda, MD*

## John C. Fletcher
*University of Virginia, Charlottesville, VA*

## James M. Humber
*Georgia State University, Atlanta, GA*

**Humana Press** Totowa, New Jersey

© 2003 Humana Press Inc.
999 Riverview Drive, Suite 208
Totowa, New Jersey 07512

**www.humanapress.com**

Production Editor: Robin B. Weisberg.
Cover design by Patricia F. Cleary.

For additional copies, pricing for bulk purchases, and/or information about other Humana titles, contact Humana at the above address or at any of the following numbers: Tel.: 973-256-1699; Fax: 973-256-8341; E-mail: humana@humanapr.com or visit our website: humanapress.com

This publication is printed on acid-free paper. ∞
ANSI Z39.48-1984 (American National Standards Institute) Permanence of Paper for Printed Library Materials.

Library of Congress Cataloging-in-Publication Data

The nature and prospect of bioethics : interdisciplinary perspectives / edited by Franklin G. Miller, John C. Fletcher, James M. Humber.
        p. cm.
    Includes bibliographical references and index.
    ISBN 0-89603-709-6 (alk paper); 1-59259-370-4 (e-book)
      1. Medical ethics--Methodology. 2. Bioethics--Methodology. I. Miller, Franklin G. II. Fletcher, John C. III. Humber, James M.

R725.5 .N385 2003
174'.957--dc21                                                      2002192211

# Preface

Most areas of intellectual pursuit, such as theology, philosophy, history, and medicine, have been the object of study for centuries. As a result, the subject matters and methodologies of these disciplines are relatively clear, and students wishing to pursue careers in these fields not only know what they will be studying when they seek their degrees, but also what they will be doing once they graduate and begin practicing their professions. Bioethics differs from these traditional areas of study. For one thing, it is only about 35 years old. For another, it is, by its very nature, interdisciplinary in character. That is to say, its practitioners come from diverse backgrounds (e.g., history, philosophy, literature, theology, medicine) and it is primarily by virtue of the communication between and among individuals from these backgrounds that the discipline has flourished and come to be recognized as a distinct field of study. Still, many students, and even some of those who have graduate degrees in the specific disciplines that contribute to bioethics, remain unclear just what "bioethics" means, what bioethicists do, how one prepares for practice in the field, and the value of the profession.

*The Nature and Prospect of Bioethics: Interdisciplinary Perspectives* seeks to provide readers with critical information of this sort. In order to accomplish this goal, the editors have carefully selected the authors who are contributors to the volume. When choosing authors, the editors used two criteria: (1) the author's prominence in his or her particular field of study, e.g., philosophy or history and (2) the reputation of the author as a bioethicist. All of the particular disciplines that are represented in the text have either contributed significantly to, or been closely associated with, the literature of bioethics and the education and professional work of bioethicists. Specific questions addressed by the various contributing authors include the following: How has the author's discipline contributed to the development of bioethics? What has been the impact of bioethics on the health care delivery system in general and the author's field of study in particular? What are the most significant current and future issues in the intersection between the author's discipline and bioethics? What issues or perspectives have been neglected? What is the state of the art of interdisciplinary scholarship, education, and service in bioethics? From the perspective of the author's discipline, what are the most significant strengths and weaknesses in the current state of bioethics? What basic knowledge and skills connected with the discipline should be acquired to obtain competence in bioethics? What methods or theoretical approaches related to the author's discipline are most promising for the future development of bioethics?

Taken as a whole, *The Nature and Prospect of Bioethics: Interdisciplinary Perspectives* seeks to achieve four general goals: (1) To explore the roots of bioethics in those disciplines that have principally informed its subject matter, (2) to illustrate how bioethics' present and future flourishing depends on its being nourished by the insights and methods that derive from those varied sources of scholarship, (3) to demonstrate the value of bioethics as a profession, and (4) to indicate the directions in which future scholarship in bioethics will most likely proceed.

One incontestable fact about any multiauthored text is that its value is almost exclusively dependent upon the ability, diligence, conscientiousness, and trustworthiness of its various contributing authors. Recognizing this fact, we would like to express our thanks to the authors whose works are included in this text. All have selflessly taken time from their busy schedules to help in the production of a volume that is intended primarily for use as an educational tool. This is a public service for which the authors should be thanked, not only by the editors, but also by other members of their professions, practitioners of bioethics, and the public at large.

*Franklin G. Miller*
*John C. Fletcher*
*James M. Humber*

# Contents

# Contributors

JOHN D. ARRAS (Philosophy): *Corcoran Department of Philosophy, University of Virginia, Charlottesville, VA*

PATRICIA BENNER (Nursing): *Department of Social and Behavioral Sciences, School of Nursing, University of California, San Francisco, CA*

HOWARD BRODY (Medicine): *Department of Family Practice, Michigan State University, Clinical Center, East Lansing, MI*

JAMES F. CHILDRESS (Religion): *Institute for Practical Ethics, University of Virginia, Charlottesville, VA*

JOHN C. FLETCHER (Religious Ethics): *Emeritus Professor of Biomedical Ethics, Division of Continuing Education, University of Virginia, Charlottesville, VA*

JAMES M. HUMBER (Philosophy): *Department of Philosophy, Georgia State University, Atlanta, GA*

ERIC M. MESLIN (Policy Analysis): *Indiana University Center for Ethics, Indiana University School of Medicine, Indianapolis, IN*

FRANKLIN G. MILLER (Philosophy): *National Institutes for Health, Bethesda, MD*

KATHRYN MONTGOMERY (Literature/Literary Studies): *Medical Ethics and Humanities Program, The Feinberg School of Medicine, Northwestern University, Chicago, IL*

M. L. TINA STEVENS (History): *Department of History, San Francisco State University, San Francisco, CA*

# 1

# The Owl and the Caduceus

## Does Bioethics Need Philosophy?

### John D. Arras

What has been the contribution of philosophy to the emerging interdisciplinary field of bioethics? Although the news might come as a shock to philosophers accustomed to their lowly and marginal status within American intellectual life, a growing chorus of skeptical casuists, feminists, social scientists, and narrativists have come to the conclusion that philosophy's role in bioethics has been both dominant and disconcerting. We hear that philosophy, especially in the guise of its Anglo-American analytic wing, has largely dominated the field for the past thirty years, bequeathing to it a distinctive language, method, and agenda. Although some of this criticism has the distinct appearance and flavor of "sour grapes" ("Hey, what about *us*!"), the charge that philosophy has played a dominant role in the formation of contemporary bioethics seems descriptively correct and well nigh undeniable. Ever since the emergence of the mantra of "autonomy, beneficence, and justice" from the primeval soup of rival moral theologies in the 1970s, the language of bioethics has been largely that of contemporary moral philosophy *(1)*. We tend to

From: *The Nature and Prospect of Bioethics: Interdisciplinary Perspectives*
Edited by: F. G. Miller, J. C. Fletcher, and J. M. Humber
© Humana Press Inc., Totowa, NJ

frame problems in the language of conflicting duties, rights, virtues, and moral principles. We wonder about the "moral status" of embryos and the brain dead, and we debate the significance and scope of the right to reproductive liberty against the backdrop of the "harm principle." We ponder the definition of genetic health and disease, and invoke different theories of justice in controversies over access to health care. Indeed, many of our debates turn on such conceptual niceties as the boundaries of coercion, competence, and personal identity.

All this seems like a distinctly philosophical enterprise. It is no great wonder, then, that so many of the leading figures in the field turn out to be card-carrying philosophers. Yet, despite the fact that philosophy's influence on the field has been pervasive, the nature of the relationship between philosophy and bioethics remains problematic. Skepticism about the value of philosophy for bioethics has emerged on at least three distinct fronts. First, a growing chorus of social scientists has conceded the dominant role of analytic philosophy within bioethics while lamenting that this influence has been largely baleful *(2)*. Second, many practitioners of bioethics make the case that they can go about their usual business of applying principles or comparing cases without having to invoke high-level philosophical theorizing. Finally, some philosophers have contended in various ways that standard brand analytical ethical theorizing is incapable of generating answers to bioethical problems. In this chapter I intend to focus on the latter two challenges—namely, the claims that bioethics can and should be largely independent of philosophy, and that philosophy is itself incapable of providing genuine help to bioethics. En route, I will argue that some of the criticisms of philosophy's influence (or lack thereof) on bioethics turn out, on closer inspection, merely to be criticisms of one particular theoretical approach to ethics, and not at all a refutation of the value of philosophy itself. I accordingly contend that although philosophical thinking is but one distinct strand within the rich interdisciplinary tapestry of bioethics, it is a crucial and ineliminable element of any bioethics that would claim to be truly reflective and critical. In closing, I suggest some ways in which bioethics may have influenced philosophy.

## Varieties of Bioethics and Philosophical Work

Does bioethics need philosophy? Can philosophy guide decision making in bioethics? At first glance, these might seem like very straightforward, factual questions, but they are not. In the first place, "bioethics" is not a monolithic entity or activity. There are at least three different kinds of bioethical work, each of which may well have a different relationship to philosophy:

1. There is *clinical bioethics*, which amounts to the deployment of bioethical concepts, values, and methods within the domain of the hospital or clinic. The paradigmatic activity of clinical bioethics is the ethics consult, in which perplexed or worried physicians, nurses, social workers, patients, or their family members call on an ethicist (among others) for assistance in resolving an actual case. These case discussions take place in real time and they are anything but hypothetical. While those who discuss bioethics in an academic context can afford to reach the end of the hour in a state of perplexed indeterminacy, the clinical ethicist is acutely aware that the bedside is not a seminar room and that a decision must be reached. Although a competent clinical ethicist has no doubt read a good deal of philosophy during his or her academic training, and although his or her approach to clinical problem solving might well exhibit some dependency on the skills of philosophical analysis, the vast bulk of the work in clinical consultation might best be described as a kind of medical ethical dispute mediation. To be sure, the content of these discussions often

revolves around philosophically charged subjects, such as informed consent, competency, the right to refuse life-sustaining treatments, and so on, but the discussions themselves are rarely explicitly philosophical.

2.  There is *policy-oriented bioethics*. In contrast to the clinical ethicist, who is concerned with the fate of individual patients, the bioethicist-*cum*-policy analyst is called on to assist in the formulation of policies that will affect large numbers of people. Such policy discussions can take place at the level of individual hospitals or health systems, where administrators, medical and nursing staff, and bioethicists debate, for example, the merits of competing policies on medical futility; or they can take place in the more rarified atmosphere of various state and national commissions charged with formulating policy on topics such as cloning, access to health care, or assisted suicide. Although such commissions operate at much higher levels of generality than the clinical ethicist in the trenches, both of these kinds of bioethical activity tend to be intensely practical and result-oriented. The clinical ethicist will usually be wary of invoking philosophical theory because his or her interlocutors usually have neither the time nor the inclination to discuss matters on this level, whereas the bioethicist on the national commission will soon realize the impossibility of forging a consensus with his or her peers on the basis of philosophical theory.

3.  Finally, at the other end of the practice–theory spectrum, there is *bioethics as an academic pursuit*, a variant unhindered by the resolutely practical constraints of the clinic and commission. The academic is free to think as deeply or to soar as high into the theoretical empyrean as he or she wishes. Unlike the clinical ethicist, the academic is unhindered by time constraints, medical custom, law, the need to reach closure, and even (some might say) common sense. The seminar lasts all semester, and it might be

just as well to leave one's students even more confused at the end than they were at the beginning. And unlike the bioethicist-*cum*-policy analyst, the academic doesn't have to worry about finding a common language, or bending to the necessities imposed by pluralism or sponsoring agencies of government. It is within this academic domain that the relationship between philosophy and bioethics will tend to be most explicit and most welcome, although even here bioethicists will need to be responsive to some of the above constraints should they desire eventually to have some influence on public policy.

A second factor that makes the present inquiry so complex and daunting is the multiple meanings of the terms "philosophy" and "ethical theory." Many critics who have questioned the need for philosophy within bioethics have tended to identify philosophy with one of the standard brand ethical theories, such as utilitarianism or Kantianism. According to this conception, the field of philosophy serves primarily as a repository of well-developed ethical theories—i.e., sets of reasons and interconnected arguments derived from a small number of fundamental principles, explicitly and systematically articulated, with some degree of abstractness and generality, that yield directions for ethical practice *(3)*. The bioethicist's job within this scenario is presumably to select the "best" theory available and then to "apply" this theory to the problems and facts at hand. However, this conception of philosophy's contribution to practical ethics is deeply problematic, especially for those involved in clinical and policy work. This is not, however, the only kind of contribution that philosophy can make to bioethics. As I observe in subsequent sections, philosophers engage in various kinds of work that can enrich bioethical inquiry, including:

- The logical criticism of arguments

- Conceptual analysis

- Developing theories of limited scope and application

- Applying metaphysical theories to issues in practical ethics

- Metaethical inquiries into methodology in bioethics

One way to proceed from here would be to attempt the laborious task of aligning these various senses of bioethics on one axis of a grid, lining up the various functions of philosophy on the other axis, and then filling in all the resulting squares with ruminations (just to pick a couple of squares at random) about how conceptual analysis has enriched clinical bioethics, or how metaethics might relate to policy-oriented bioethics. Because this route promises to be mind-numbingly tedious, I begin, instead, with a brief discussion of one highly valuable, yet completely uncontroversial, contribution of philosophy to bioethics: i.e., the critical analysis of arguments. I then take a closer look at the two challenges mentioned above, to the relevance of philosophy for bioethics—namely, the claimed independence of bioethics from ethical theory, and the purported insufficiencies of philosophical theory for concrete problem solving. Following my assessment of these challenges, I explore some alternative conceptions of (and expectations for) philosophical work within bioethics.

## Bioethicists as Logical Traffic Cops

Although many people, including some philosophers, dispute the ability of philosophy to generate plausible and useful positive theories, no one doubts the ability of philosophers to spot fuzzy thinking and demolish bad arguments. Weaned on the rigorous study of logic from the beginning of their professional training, philosophers are exceptionally good at spotting logical fallacies, disambiguating the meaning of propositions, criticizing definitions, mapping the logical structure of arguments, and pinpointing their missing premises and flawed inferences. This mode of philosophy has generated some extremely important work in bioethics. Although this kind of careful logical brush-clearing can be found in just about any serious article picked randomly from a self-respecting bioethics journal, I highlight just one well-known example here from the work of the President's Commission's landmark study, *Deciding to Forego Life-Sustaining Treatment (4)*.

Published in 1983 when public discussion of these difficult issues was just getting off the ground, the Commission's report targeted several important distinctions that had entered into common parlance in the news media, court decisions, and respected periodicals. Although these distinctions (e.g., between withdrawing and withholding medical treatments, ordinary vs extraordinary treatments, and acting vs omitting) enjoyed wide currency at that time, the Commission's conceptual and ethical analysis of them was devastating and definitive. In each case, the Commission showed convincingly, first, how the purported distinction was conceptually murky. For example, the distinction between actions and omissions was shown to be fatally unclear when deployed in this connection. Is a decision to discontinue ventilator therapy a "mere" omission, in which case it would supposedly be morally permissible; or is it an action of sorts (one is physically *pulling* the plug or *flicking a* switch), in which case it would presumably be morally illicit? According to the Commission's analysis, such decisions could quite easily fall under either description, thus rendering such descriptions singularly unhelpful in moral deliberation.

The Commission showed, secondly, that each of the purported distinctions tended to focus attention on irrelevant moral considerations. Thus, in connection with its discussion of the ordinary vs extraordinary treatment distinction, the Commission argued convincingly that such distinctions tended to focus our attention on the intrinsic characteristics of various treatment modalities (e.g., asking whether a ventilator is "experimental" or "high-tech") at the expense of a more ethically meaningful inquiry into the expected impact of the proposed treatment on the individual patient's condition. In some instances, the Commission helpfully pointed out that some of these distinctions (e.g., between withholding and withdrawing treatments) could actually result in *harming* patients by tempting physicians to forgo initial trials of therapy that could eventually prove beneficial or even life-saving.

Although the Commission's valuable report on forgoing life-sustaining treatments was the result of an extraordinary interdisciplinary confluence of staff physicians, legal scholars, social scientists, philosophers, and politically appointed commissioners, the debunking of these muddled and mischievous distinctions was uniquely the handiwork of the staff philosopher, Dan Brock of Brown University. The great virtue of this sort of work is that it clears from our path a lot of unruly conceptual brush, so that we can concentrate on more morally relevant matters. Its obvious shortcoming lies in its negativity. Logic policing tells us which distinctions and arguments we shouldn't make, but it doesn't tell us what we *should* be thinking and doing.

## Two Challenges to the Relevance of Philosophy

### *The Independence of Bioethics*

As several observers of the bioethical scene have recently noted, the methodological strife amongst the proponents of principlism, casuistry, and narrative ethics seems to be giving way to a common recognition that each of these approaches contributes something of value to an emerging consensus on method within practical ethics *(5)*. Although individual commentators may emphasize one approach over the others, just about any bioethical presentation will encompass: (1) an invocation of moral principles and an attempt to specify, weigh, and balance them; (2) analogical comparisons of the present case to other cases for the purpose of ascertaining the strength of various principles in different factual contexts; and (3) richly detailed narratives of the problem and the involved protagonists designed to highlight the morally relevant matters at stake, including their relations to one another, their motivations and emotional responses, and the meaning of various outcomes for each participant.

Conspicuously absent from this inventory of bioethical tools is any reference to high-level philosophy or ethical theory in the service of concrete problem solving. The practitioners of principlism, casuistry, and narrative ethics all heartily endorse the claim that these relatively low-level methodological techniques suffice for the identification and resolution of most bioethical problems. There is simply no need, they argue, to invoke high-level philosophical theory. Indeed, they assert that appeals to philosophical theory actually serve to hinder the search for solutions among well-motivated ordinary people confronting difficult moral choices in real life *(6)*. Thus, the principlist asserts that we can make do with various middle-level principles, such as autonomy, beneficence, non-maleficence, and justice. Although we can, and do, disagree vehemently at the ultimate level of theoretical justification, the principlist asserts that these disagreements often wither away at the middle level of principles and rules (e.g., "respect the autonomy of patients," "honor patients' right to confidentiality," and "do not subject patients to needless risk."). While some partisans of ethical theory might insist that the inevitable conflicts between these middle-level principles and rules must ultimately be resolved by appeals to a theory complete with its own priority rules, principlists are usually content to work out such conflicts *in medias res* with the aid of nothing more than a rich appreciation of context and sound judgment. This kind of approach has actually left some partisans of philosophical theory deeply dissatisfied with principlism. According to these critics, principlism's unwillingness to organize its different mid-level norms by means of theoretically derived priority rules results in its inability to resolve conflicts among principles and, therefore, to guide action successfully. These partisans of theory dismiss the norms of principlism as "mere chapter headings" *(7)*.

The advocates of casuistry in bioethics have likewise argued for the self-sufficiency of reasoning by appeal to paradigm cases

and analogy. While conceding the existence of some disparate uses for higher level moral philosophy, the partisans of
casuistry have claimed that bioethicists can, for the most
part, get along quite well without recourse to ethical theory
*(8)*. They contend *inter alia* that reasoning by paradigm and
analogy is a powerful engine of thought happily geared, in
stark contrast with moral theory, toward the resolution of
practical disputes. They also contend that consensus can be
achieved in the midst of pluralism, as one proponent puts it,
through "incompletely theorized agreements" forged by analogical thinking *(9)*. Finally, they hold that an emphasis on
high-level theory will lead only to irresolution and moral
fragmentation *(10)*.

For their part, the advocates of narrative contend that their
emphasis on the particulars of the patient's story serves to illuminate the morally salient facts of a case, including the life-trajec-
tory of the patient and her family, the web of relationships
enmeshing all the major characters (including the caregivers), and
whatever role-based duties that may apply deriving from the guiding stories of a family, social group, or nation *(11)*. Narrativists
as a group are decidedly more particularistic than either their
principlist or casuist brethren, and are usually content to under-
score the morally salient aspects of a case without explicitly
invoking principles or rules. They tend to regard the invocation
of moral theory as a distracting abstraction that obscures what's
really going on in a case.

Thus, all three of these complementary methods proclaim
the independence of practical ethics from philosophical
theory. As resolutely practical methods of thought, they appeal
(both individually and in combination) to medical practitioners and ordinary people caught up in the search for solutions
to common problems. Especially in the contexts of clinical and
policy-oriented bioethics, appeals to high-level philosophical
theory may well turn out to be both unnecessary and counter-
productive.

## The Limits of Philosophy for Practical Ethics

Momentarily bracketing claims just made for the independence and sufficiency of principlism, casuistry, and narrative for the resolution of bioethical disputes, let us turn to some recent critiques regarding the adequacy of analytical moral philosophy for this purpose. Interestingly, these critiques have come, not from the quarter of knuckle-dragging, pragmatic physicians, but rather from distinguished moral and political philosophers. Thus, Will Kymlicka contends, on the basis of his own experience on a Canadian reproductive ethics commission, that "moral philosophy" is of decidedly little use in the discussion and resolution of public policy issues *(12)*. Although Kymlicka believes that such commissions should definitely take *morality* seriously, he doubts that they should take *moral philosophy* seriously.

Kymlicka distinguishes between two different philosophical agendas with regard to practical ethics: (1) a "modest" approach, which amounts to logic and argument analysis; and (2) a more "ambitious" view requiring the adoption and application of a normative ethical theory. He gives the modest view short shrift, noting that logical consistency, coherence, and conceptual clarity aren't enough. As he puts it, "arguments can be clear and convincing yet be morally bankrupt" insofar as they ignore the interests of those affected by them *(13)*. Bioethics will require more from philosophy than mere technical proficiency.

Kymlicka's rejection of the ambitious view is more complex, tracking each of the steps required to identify the leading ethical theories, selecting the best one, and then applying it to a practical problem. First, he offers a short list of ethical "positions," including: utilitarianism, deontology, contractualism, natural law, and the ethics of care. A problem of interpretation immediately arises whereby we often have difficulty pigeonholing individual philosophers into one or the other of these theoretical niches. Second, there is the even more daunting problem of vindicating a single ethical theory as the best for purposes of public policy. Kymlicka makes the obvious point here that, so long as the membership of a public commission is even remotely representative of the larger society from which it is drawn, it will be impossible to achieve consensus on a single, correct ethical theory. Philosophers have been disagreeing over the conflicting truth claims of such all-encompassing, comprehensive theories for centuries, and we have no reason to expect the members of a public commission to succeed in achieving a consensus where the allegedly best and brightest have failed. Even supposing, *per impossibile*, that we could agree on the best theory, there would then be the problem of how best to characterize that theory. Here Kymlicka observes that philosophers pledging allegiance to the same basic theory may still disagree about the best interpretation of that theory, just as act utilitarians disagree with rule utilitarians, or Thomistic partisans of natural law disagree with Lockians. And finally, assuming (again *per impossibile*) that everyone could agree on the best interpretation of the best theory, the commissioners would face serious difficulties in applying their favored theory to the particulars of concrete policy decisions. Thus, utilitarians may, and do, disagree amongst themselves over the legal permissibility of surrogate parenting, and Kantians disagree about the morality of assisted suicide and euthanasia. Hence it is unrealistic, Kymlicka concludes, to expect a public ethics commission to go about its business by adopting and then applying an ethical theory.

Having dispensed with both the modest and ambitious views of philosophy's usefulness to practical ethics, Kymlicka advances the proposal that public ethics commissions should eschew ethical theory in favor of the much more modest agenda of identifying those people who will be most affected by a proposed policy, and then attempting to promote (or at least not frustrate) their legitimate interests. He adds that this emphasis on the interests of the various stakeholders can and should be supplemented by some equally modest mid-level principles that enjoy widespread public recognition and support—e.g., the principles of autonomy, accountability, respect for human life, equality, careful shepherding of resources, the noncommercialization of reproduction, and the protection of the vulnerable *(14)*. Thus, instead of emerging from the cave in search of the form of the Good, the members of public ethics commissions should attempt to base their policy recommendations on sensitive estimates of the affected interests and on widely accepted values within a pluralistic society. Kymlicka concludes with the observation that neither of these more appropriate tasks requires the knowledge or skills of an ethical theorist. Indeed, he contends that philosophical sophistication might make it more difficult for commissioners to exhibit the necessary moral sensitivity to affected interests and public values.

## How Independent is Practical Ethics from Philosophy?

I begin my critical assessment of these two challenges to the usefulness of philosophy for bioethics with the "independence thesis"— i.e., the claim of assorted principlists, casuists, and narrativists to the effect that they have little, if any, need for higher level philosophizing. In order to secure our bearings at the beginning of this reply, it might be useful to list some particular examples of common moral problems encountered by bioethicists in clinical, policy-oriented, and academic settings. Doing so will perhaps give us a bit more traction as we attempt to assess the usefulness of philosophy for bioethics. The following is a list, off the top of my head, of some typical (but by no means fully representative) examples:

- Is Ms. Smith, an elderly and mildly demented woman, capable of deciding whether to undergo a somewhat risky medical procedure?

- Should caregivers accede to a family's request to pull a feeding tube from their permanently vegetative matriarch?

• Is the withdrawal of life-sustaining nutrition and hydration the equivalent of murder?

• What should be the appropriate standard for diagnosing death in New York?

• Is it morally permissible to destroy embryos in order to secure stem cells for the development of future therapeutics?

• Is there a right to health care?

• How should we determine priorities for allocating scarce health care resources?

• Is cost-effectiveness analysis a just method of rationing health care?

• Is society responsible not only for treating disease, but also for enhancing various "normal" (i.e., nondiseased) traits, such as short stature or low intelligence?

• Can someone be harmed by being born? Do parents act wrongly by knowingly bringing children into existence with serious disabilities?

• Are various modalities of assisted reproductive technologies "unnatural" and, if so, in what sense? Should this matter in policy deliberations?

• Is there a moral right to assisted suicide?

• Do placebo-controlled, randomized-controlled trials imperil subjects' rights?

• Do incentives coerce poor women to use birth control?

• Do researchers from wealthy nations exploit subjects in poor countries?

• Do the citizens of developed countries have a duty to assist those in developing countries to obtain food and health care?

Turning now to the independence thesis, the question is whether and to what extent principlists, casuists, and narrativists can avoid dependence on moral philosophy as they go about their bioethical business. This thesis, I believe, harbors a very large core of good sense. It is undeniably true that those working in bioethics, and especially in more clinically and policy-oriented quarters, can get along quite well most of the time without invoking higher level appeals to philosophical ethics and especially to comprehensive ethical theories. Beginning with a thick narrative description of the case and its cast of characters, and then moving on to judgments based on mid-level moral principles and careful case comparisons, the practical bioethicist can go a long way without theory. Thus, the bioethicist can work his or her way through the tube-feeding issue mentioned in the above list with some careful analogical comparisons of the instant case with clearly permissible and impermissible instances of withdrawing life-sustaining treatments. If the present case more closely resembles the permissible cases, then he or she may justifiably conclude that withdrawal of tube feeding is morally justified in that instance.

This kind of practical moral reasoning has two major advantages. First, it can meaningfully engage the participation of ordinary folk (i.e., non-philosophers on ethics commissions, health care professionals, patients, their families, etc.) who have neither the time nor the inclination to develop an expertise in moral theory. Second, it allows for a rational discussion of issues and cases without requiring an appeal to a fully articulated and comprehensive moral theory *(15)*. Because the raw materials of this level of moral thinking—namely, mid-level principles, case comparison, and narrative description—also serve as the starting points for higher level moral theory building, we can engage in moral thinking on this level with the assurance that we are indeed engaged in a rational (albeit not fully theorized) enterprise. Indeed, we often feel more confident in our judgments arrived at on this more concrete level than we do about the deliverances of moral theory, and we sometimes reject theories that contradict our most firmly held intuitions (or "considered judgments") about cases and mid-level generalizations *(16)*.

Having conceded this core of good sense to the independence thesis, we return to our list of issues and notice immediately that some of these questions pose rather deep questions of moral and political philosophy. For example, a state commission contemplating a revised policy on the criteria of death will have to ask and answer some difficult questions about the meaning of death and the proper concept of death. A federal commission charged with forging policy on stem-cell research will have to confront the notoriously complex philosophical question about the moral status of human embryos. An inquiry into the right to health care would have to grapple at some length with the nature of rights and how rights claims might be justified, as well as with the substantive arguments relating to claims to health care. An investigation into the moral acceptability of cost-effective analysis would necessarily entail a philosophical assessment of the moral theory of utilitarianism. And finally, a determination of the responsibilities of parents for choosing to have a child who will be born with serious deficits will require a prolonged detour into the heart of Derek Parfit's so-called "non-identity problem" *(17)*.

All of these examples indicate that, as working bioethicists go about their daily rounds, they will occasionally run into problems that require what Ronald Dworkin has called a "justificatory ascent" *(18)*. As practical people, bioethicists typically work, like Dworkin's lawyers and judges, from the "inside-out"; i.e., we begin with concrete problems that come with the territory of doing ethics in a clinical setting or public policy council. We have often encountered similar problems in the past, and can usually solve the new problem by using arguments ready-to-hand and extrapolating from our previous findings in related cases or policy discussions. Although we often have to be painstakingly careful in deciding what is the right thing to do, the scope of our inquiry will usually be severely limited by time constraints, medical and social custom, law, and the facts served up by the case at hand. Occasionally, however, our quotidian methods prove insufficient, and we find ourselves drawn upward in a justificatory ascent toward a more theoretical response.

Such an ascent may be motivated in several ways. For example, we may encounter a problem that is so fundamentally philosophical that it cannot be adequately confronted with anything less than a full-scale theoretical argument. I would contend that the problem of embryo and stem-cell research fits this description. Efforts to avoid the philosophical puzzle of the embryo's moral status, or to pretend that one is not answering it, inevitably strike the reader and the public as evasive and implausible *(19)*.

Another impetus to justificatory ascent is provided by conflicting principles that threaten the coherence of our overall approach to ethical problems. We might, for example, come to a point where we are fairly confident that a certain interpretation and specification of a particular moral principle provides us with the best justification for a public policy. (Imagine, for example, that we are members of a panel that is prepared to legitimize surrogate parenting via the principles of autonomy and reproductive liberty.) But if, as Dworkin suggests, we raise our eyes a bit from the particular considerations that seem to us most on point in the instant case, and look at neighboring areas of concern in either law or bioethics, we may realize that a principle we have downplayed in this case has weighed very heavily in other areas of concern to us *(20)*. (For example, our hypothetical reproductive ethics panel may have downplayed concerns about the commodification of human life in endorsing a lenient policy on surrogate parenthood, only to be reminded that the principle of non-commodification has played a pivotal role in public policy in the area of transplantation ethics.) Once we are confronted with these kinds of conflict among principles (e.g., between respect for reproductive freedom and non-commodification), we have to raise our analysis to a higher level and ask why reproductive autonomy should outweigh non-commodification in this instance, but not in others. Posing and addressing such questions puts us well on the way to developing, at least, a localized theory of reproductive autonomy and its limits.

Finally, justificatory ascent can be forced on us "from above." For instance, we go about our usual practice, dealing with cases in the routine way with the usual materials, when suddenly the status quo is challenged by what Dworkin refers to as a "new and potentially revolutionary attack from a higher level" *(21)*. Within the field of bioethics, a pertinent example of such a revolutionary threat is provided by the feminist critique of our standard ways of doing business across a whole host of fronts, including research ethics, the physician–patient relationship, access to health care resources, targets for health-related development efforts in foreign countries, and (obviously) reproductive ethics *(22)*. Often the point of such theoretically inspired feminist criticism is that the well-trodden grooves of our habitual analogical thinking have led us in the wrong direction. Left to their own devices, methods designed to hug the ground and deliver "incompletely theorized agreements" may lack the theoretical resources necessary to identify flawed lines of analogical thinking. We become so inured to standard "middle level" ways of thinking and doing that we fail to see the injustice embedded in them *(23)*.

The goal of coherence in our moral views that animates Dworkin's story of justificatory ascent suggests an entirely different slant on the question of philosophy's relationship to bioethics. What Dworkin refers to as justificatory ascent—i.e., bringing the principles operative in one ethical or legal domain into a harmonious relationship with sets of principles operative in neighboring corners of our moral and legal universe—bears a striking resemblance to what moral and bioethical theorists have termed the search for "reflective equilibrium" *(24)*. This term refers to a dynamic process in which we advance reasons for a certain principle or theoretical position, comparing them against the provisionally "fixed points" of our most firmly held moral intuitions. We then move in two directions, both amending our principles and theories to fit our firmest intuitions, and judging our (less firm) intuitions from the vantage point of more confidently held principles and theory. Although Rawls relied on this method to identify and justify his two fundamental principles of justice, many bioethicists have embraced reflective equilibrium as a method for rendering coherent, and thereby justifying, all manner of moral judgments bearing on cases, policies, principles, and higher level theories *(25,26)*.

Importantly, many of these methodologically oriented bioethicists have claimed that reflective equilibrium, rather than appeals to high-level ethical theory, is the most preferred method for justifying moral judgments *(27)*. Indeed, some go one step farther, arguing that our best account of "theory" is precisely nothing other than the full set of our most confidently held moral intuitions, mid-level principles, and background theories (e.g., of human nature, social structure, etc.) all in reflective equilibrium with one another *(28)*. Clearly, ordinary mortals working in the fields of practical ethics cannot actually aspire to achieve such a comprehensive ordering of our moral commitments on all these different levels, but we can at least hope to achieve coherence in more limited domains of inquiry.

Were we to accept this approach to moral justification and to the very meaning of philosophical theory, it would cast our present inquiry in an entirely new light. Instead of viewing practical ethics as an activity that is largely aloof from more theoretical concerns and defined by entirely different methods than those of moral philosophy, this conception of "theory" as reflective equilibrium yields the view that the methods of practical ethics and moral philosophy are identical *(29)*. In each of these we work with the moral data at hand, struggling to align our intuitive convictions, principles, and background theoretical commitments into some sort of coherent package. Thus, in this view, the gap dividing practical ethics from moral philosophy has been bridged by a common method of moral reasoning and moral justification. The clinical ethicist working in a neonatal intensive care nursery, the policy-oriented bioethicist working for a state or national commission, and the moral philosopher in his or her study are all operating, according to this understanding of justification, with the same basic method of moral reasoning. True, each works in a different domain and with greater or lesser degrees of abstraction, which will affect the region or scope of moral experience within which they will seek to discern coherence. However, all are engaged in a common enterprise of moral philosophizing.

If we accept this picture of moral justification as residing in reflective equilibrium, then practical ethics will be saturated with theory rather than being independent from it. Once we understand moral philosophizing to be the quest for greater justificatory coherence among our disparate moral commitments, then "theorizing" will be tantamount to heeding the call of justificatory ascent. We would theorize whenever we "lifted up our eyes" to confront manifestly philosophical problems (e.g., the moral status of embryos), to acknowledge a lack of coherence in different regions of our moral experience, or to respond to a revolutionary attack from a higher level (e.g., feminism). This is not to refute the thesis, as previously understood, about the independence of practical ethics from moral philosophy. That thesis understood justification in moral philosophy to be an essentially deductivist enterprise, in which one would first identify the "correct" moral theory (i.e., utilitarianism, Kantianism, etc.), and then deduce proper conclusions from it regarding states of affairs. If we understand moral philosophy in this traditional way, then the independence thesis remains largely true. If, however, we understand moral philosophy to be the quest for reflective equilibrium, then a radically different picture emerges of the relationship between practical ethics and moral philosophy, one in which they are unified by a common method.

## How Limited is Moral Philosophy?

Will Kymlicka and others claim that moral philosophy cannot fruitfully inform the practical judgments of public ethics commissions. A similar skepticism might apply for analogous reasons to any bioethicist intent on influencing public policy. I respond to this second major challenge to the relevance of philosophy to practical ethics with yet another concession. Insofar as these critics of moral philosophy understand it to be equivalent to traditional deductivist uses of moral theory (e.g., "What would Kant say about assisted suicide?"), they are to a large extent on the mark *(30)*. They are correct in claiming that there is no usable consensus as to the "best" theory, and that each theory is itself a battleground of rival interpretations capable of issuing incompatible moral solutions to the same problems. We might add that ethical theory so understood tends to be perniciously reductionist, paving over the rich diversity of moral experience to make the world safe for one or two carefully chosen principles. Expecting a public ethics commission to adopt such a theory would not only be an exercise in futility (because consensus will never be achieved), but would also be profoundly anti-democratic. Even assuming (*per impossibile*) that a consensus at the level of traditional theory (e.g., in favor of utilitarianism) could somehow be rigged up among society's elites, those citizens whose reasonable moral views remained at odds with this putative consensus would have a right to complain that their government had imposed a vision of the good upon them.

Does the essential rightness of the attack on the relevance of ethical theory also demonstrate the impotence or irrelevance of moral philosophy generally for practical ethics? I argue that it does not. Beginning with Kymlicka's proposed evasion of moral philosophy in favor of an interest-based analysis of moral concerns, recall that Kymlicka intends to take "morality" seriously while avoiding "moral philosophy" (at least for purposes of public ethics commissions). Taking morality seriously amounts to assessing the projected impacts of various policy proposals involving new technologies upon the interests of all concerned. This enterprise crucially relies, not on high-level philosophical skills, but rather on the ability to empathize with other people, to walk around in their shoes in order to experience how their lives might be bettered or blighted by the new technology.

Although I agree with Kymlicka that attentiveness to the interests of involved stakeholders is a crucial task of any public ethics commission, I do not believe that it is sufficient to yield sound ethical advice. The first problem with this position is that it simply ignores the testimony of our list of standard issues in bioethics. Many of the problems mentioned in this list cannot (or so I would argue) be rendered more tractable by being re-described exclusively in the language of interests. One might be able to effect such a re-description for the question involving the right to health care, because rights are often explained and justified in terms of interests. However, Kymlicka will be less able to deal exclusively in these terms with, for example, the question of embryo research, wrongful disability, the putative coerciveness of birth-control incentives, and establishing criteria for death.

To develop just one of these counter examples briefly: how might one attempt to answer the question about allegedly "wrongful disability" exclusively in the language of interests? The problem here is that, *ex hypothesi*, certain children have no other choice but to be born with certain handicaps or not at all. Assuming that their lives are not so grim as to be deemed "wrongful"— i.e., assuming they would not be "better off" never having been born—we must conclude that it is in their interest to be alive. If asked, they might respond, "Yes, I've had a hard life with all these disabilities, countless surgeries, and constant discomfort, but still I'm glad to be here." The question, however, is whether it is morally responsible for parents to knowingly bring children into the world in such a harmed condition. Even if such children have not been "harmed," in the sense that being born did not make them worse off than they otherwise would have been, and even if we cannot therefore label them as "victims," it still might make good moral sense to conclude that parents do wrong or act irresponsibly by knowingly giving birth to them under such circumstances *(31)*. In a case such as this, it may well be that a moral analysis carried out exclusively in terms of interests would reach precisely the wrong result.

Another problem with Kymlicka's "moral theory-free" method of attentiveness to interests is its silence on important questions relating to how such interests will be tallied up, weighed, and compared. Take, for example, the item from our list pertaining to setting priorities in the allocation of scarce health care resources. Suppose we are members of a managed care policy council charged with the task of determining whether high cost, experimental, and last-chance therapies (e.g., bone-marrow transplant for advanced breast cancer) should be covered by the plan and, if so, which patients should be given priority to receive them. An analysis focused exclusively on interests would most likely not get us very far. *Ex hypothesi*, all of the possible candidates for such a therapy are desperately ill; thus, all have a major interest in receiving the treatment. What conclusions can be drawn from this fact? Not many. In order to advance the policy discussion, we will need to make choices freighted with theoretical implications. We might, for example, decide to maximize the number of quality adjusted life years (QALYs), which would most likely lead us to abandon last gasp bone-marrow transplants and invest our plan's money elsewhere. Or we might decide to favor those who are worse off, despite the fact that we could further more interests by investing in prenatal care or breast-cancer screening. In any case, merely tallying up the interests of various stakeholders will not provide us with a guide to action.

Finally, Kymlicka's evasion of philosophy falls short because it fails to acknowledge the fact that the policy world of bioethics is already saturated with philosophical theory (or fragments thereof), much of which is deficient. The only way to get rid of this bad philosophy is to substitute better philosophy in its place *(32)*. Importantly, however, the philosophizing that we do need not conform to the standard picture of moral philosophy as standard brand ethical theory. As we have already seen in our previous discussions of justificatory ascent and reflective equilibrium, there are more ways of doing philosophical theory than are usually contemplated by those who charge that philosophy has little to offer policy-oriented bioethics. In my closing section, I explore some other means of conducting moral philosophy and how this might enrich bioethical inquiries.

## Alternative Conceptions and Contributions of Moral Philosophy

Assuming that the critics are right, and that standard brand ethical theories will not be of any significant help to bioethics (and especially to those working in clinical ethics and policy domains), what kind of philosophy can enrich bioethics? Annette Baier draws a potentially helpful contrast between two different approaches to moral theory construction *(33)*. On the one hand, we can build by means of a "mosaic" method—i.e., by assembling a number of smaller scale projects, brick-by-brick as it were, until we have something close to a product that is both coherent and complete. Baier terms this the "weakest sense" of theory, finding that most of the moral theorizing currently done by women fits this "mosaic" mold. On the other hand, there are (what I've been calling) the standard brand ethical theories, which Baier analogizes to architectural vaults. Instead of building up a wall brick-by-brick, this kind of traditional theorizing attempts to provide a tightly systematic account of a large area of morality held together by a pivotally placed keystone (e.g., the principle of utility or the categorical imperative). In what follows, I ally myself with Baier's weaker, incrementalist conception of theory, which I believe, bears some significant resemblance to the method of reflective equilibrium.

### *Conceptual Analysis in the Service of Practical Ethics*

Philosophers have made significant contributions to the field of bioethics through the analysis of several crucially important, but highly problematic concepts. These analyses fittingly illustrate my claim that the field of bioethics has been littered with fragments of extremely bad theory, and that better theory can indeed clear our heads and direct our thought in more productive directions. Here are just three examples:

### *Benjamin Freedman's Analysis of "Equipoise" in Clinical Research*

This analysis bears on the issue of when it is ethically permissible to begin or terminate a clinical trial, which usually compares a standard therapy with some innovative drug or procedure. It has long been acknowledged in the field of research ethics that it is impermissible to begin such a trial unless all its proposed arms are in a state of "equipoise" (i.e., unless researchers are convinced that no proposed arm is superior to any of the others based on prior clinical or research experiences). This was commonly interpreted to mean that if the physician–researcher had any reason to prefer one regimen over another—based, for example, on his or her personal experience, idiosyncratic interpretation of the available data, etc.—then it would be unethical for the clinician to enter his or her patients in a trial, even if such a trial promised significant medical benefits to society. As Benjamin Freedman recognized, this is bad philosophy. In a classic paper that established a new paradigm for the research ethics field, Freedman distinguished between the above version of equipoise, which he termed "theoretical equipoise," and another version, "clinical equipoise," that placed the scientific–medical community at the center of attention *(34)*. For Freedman, the key question wasn't what individual physician–researchers happen to think, but rather whether there is still an honest disagreement within the relevant clinical community; a disagreement that a well-designed clinical trial might resolve. If Freedman's proposed distinction ultimately proves successful, it will allow physicians to enter their patients in clinical trials with a clear conscience, and without having to resort to unsatisfactory utilitarian justifications.

This distinction between theoretical and clinical equipoise has exerted significant influence, both at the local level of individual institutional review boards (IRBs), and at the national level in the reports of the National Bioethics Advisory Commission (NBAC) *(35)*. As with any paradigm-shifting piece of philosophy, Freedman's concept of clinical equipoise has occasioned a spirited debate within the field of research ethics, and many critics have doubted its ability to resolve all the problems that Freedman had hoped it would. Still, there is no denying that his distinctly philosophical–analytical contribution has enriched the field tremendously *(36)*.

### Allen Buchanan and Dan Brock on the Concept of Decision-Making Capacity

A crucially important and ubiquitous question in clinical medicine is whether a particular patient should be considered an autonomous decision maker, or whether decisions regarding the patient's treatment should be made by others on her behalf. Quite often, busy clinicians solve this problem by calling in so-called "liaison psychiatrists" to declare the patient either "competent" or "incompetent." Sometimes the mere fact that a patient declines supposedly necessary treatment is sufficient to get that patient declared incompetent; alternatively, psychiatrists often declare patients "incompetent" merely on the basis of a brief mental status examination. Again, this is bad philosophy, which Allen Buchanan and Dan Brock have sought to remedy in their complex but elegant examination of the concepts of "competence" and "incompetence" *(37)*. The high points of this conceptual analysis include the observations that these terms are "decision-relative" (i.e., they are geared to particular decisions rather than to global capacities); that competency determinations, carried out in a decision-relative manner, are "all or nothing" rather than being matters of degree *(38)*; and that standards for decision-making competence should be geared to the level of risk posed by a particular decision. Again, one can agree or disagree with the various elements of this conceptual analysis, but I contend that it has substantially raised the bar of discussion on this important topic.

### *Bruce Miller on the Concept of "Autonomy"*

A related conceptual morass has centered on the various uses of the term "autonomy" in the context of patients' decisions to refuse life-saving treatments. In the early days of the contemporary bioethics movement, some physicians, eager to show newfound respect for patients' autonomous wishes, were deeply conflicted by cases in which patients were explicitly and forcefully refusing treatments under conditions that engendered suspicions regarding the soundness or rationality of those decisions. Were physicians acting "paternalistically" (and therefore unethically) when they opposed an explicit decision of their patients? Those who thought so were, according to Bruce Miller, advocates of bad philosophy. Miller's contribution to this debate was to "unpack" the concept of "autonomy" into four distinct senses: namely, autonomy as free (uncoerced) action; as authenticity; as effective deliberation; and as moral reflection *(39)*. By showing that one could be autonomous in a weak sense (i.e., as acting freely) while being non-autonomous in a stronger and more important sense (e.g., failing to engage in effective deliberation), Miller helped physicians and bioethicists to achieve a more complex, nuanced, and morally adequate understanding of both autonomy and paternalism *(40)*.

### *Mid-Level Theory Building*

Another genre of useful bioethical work involves theorizing at the middle level. Perhaps the best and most successful example of this genre is Norman Daniels' theory of "just health care" *(41)* Daniels builds his theory on the basis of two elements: an account of what's special about health care needs (an element requiring some conceptual analysis), and a normative principle of equal opportunity. In a nutshell, Daniels argues that health care is special because of its pivotal role in achieving equal opportunity, and that this fact, together with a robust principle of equal opportunity (as opposed to merely formal opportunity), can yield a right to health care. Although Daniels is clearly attempting to extend a Rawlsian paradigm of justice to health care, he insists that one need only embrace a robust principle of equal opportunity (of whatever provenance) in order to be persuaded by this theory *(42)*. His argument thus has a distinctly hypothetical character: *if* one agrees to a mid-level principle of equal opportunity, *then* given the crucial importance of health care for opportunity, one should accept the conclusion that the members of our society have an entitlement (grounded in justice) to health care.

One advantage of Daniels' kind of theorizing is that it does not compel the reader to adopt a full-blown ethical–political philosophy in order to accept his conclusions. Given the manifest lack of consensus on these theoretical matters underscored by critics such as Kymlicka, this is a good move. Even in the absence of a foundational theory, Daniels' account of just health care can still make a significant contribution to public debates about access to health care, just as long as his chosen mid-level principle enjoys sufficiently widespread acceptance in the society at large. Clearly not everyone will agree with Daniels' principle of equal opportunity—libertarians certainly won't—so his kind of theorizing is not designed to compel universal assent *(43)*. But it should be sufficient, or so I am inclined to argue, that such a theory illuminates a wide swath of policy terrain while achieving widespread acceptance on the basis of widely shared moral convictions.

What concrete results might we expect from a theory like Daniels'? This question highlights the problematical relationship between the deliverances of moral theory (even at the middle level) and concrete decision making in clinical ethics or policy. The common complaint has traditionally been that the conclusions of theory are too abstract and remote from the details of concrete moral life to connect in a compelling way to the exigen-

cies and particularities of moral and political choice. Here every-
thing depends on the degree of specificity that we expect from a
moral theory. If we ask whether Daniels' theory can justify the
establishment of *institutions* guaranteeing access to health services as a matter of right within a larger set of institutions
designed to secure equality of opportunity, then I think the
answer is clearly yes. (Obviously, I accept his basic principle of
opportunity.) On the other hand, however, if we ask a more specific question about the availability of particular treatments or
diagnostics—e.g., does everyone have a moral right to expensive,
experimental, life-saving therapies?—then the answer is equally
clearly no. In other words, Daniels' theory of just health care can
support the establishment of *institutions* charged with providing
free access to health care for those who need it, but it cannot tell
us anything helpful about which particular expensive treatments
should be made available within that institutional framework.

The conclusion that Daniels draws from this shortcoming of
his theory, rightly to my mind, is that at some point the results of
our moral and political theorizing will have to be supplemented
with appropriately structured democratic deliberations *(44)*. A
good theory will take us part of the way—importantly, it will tell
us that we should not leave access to health care entirely to the
vagaries of the market—but it will not be able to tell us all we
need to know. Thus, in order to determine exactly how competing health care services should be prioritized, how much weight
should be given to assisting the worst off vs getting the most
"bang for the health care buck," etc., we will have to rely on
democratic politics, and not philosophy. This is just one instance
of the more general truth that philosophical theory, whether standard brand or middle level, cannot reasonably be expected to generate algorithms for practical decision making. There is no substitute
for the old-fashioned virtues of discernment and prudence.

Daniels' account of just health care is just one of a number
of important theories developed in the space between concrete
cases and full-blown traditional moral theories. Other examples

include theories of harm *(45)*, informed consent *(46)*, confidentiality and privacy *(47)*, autonomy and paternalism *(48)*, and competence *(49)*. Each of these theories is obviously of limited scope—as Baier recommends, we are indeed working brick-by-brick here—but each manages to illuminate crucial areas of moral experience in medicine. Predictably, there is no uniquely authoritative theory governing any of these topics; competing theories emphasizing different values vie for our attention. Like Daniels' theory, these small-scale efforts in theory-building are often developed on the basis of middle-level principles that theorists hope will be widely acceptable within the medical profession and the larger society *(50)*. They do not depend on the soundness or universal acceptance of any traditional ethical theory based on a small set of abstract fundamental principles (i.e., Baier's keystones), and are usually constructed at least in part from the bottom-up, on the basis of careful case comparisons. Such theories are, in my opinion, absolutely essential if we are to engage in rational discussion and debate about the nature of any of these moral phenomena (e.g., informed consent), the diverse values that serve to justify principles, rules and virtues with regard to them, as well as how conflicting values (e.g., between respect for autonomy and the desire to protect the vulnerable) should be mediated *(51)*.

## Metaphysics in the Service of Bioethics

Another kind of philosophical theorizing that might have some import for bioethics is straightforwardly metaphysical. Just as bioethics is saturated with (fragments of) moral theory, so it is also saturated with metaphysical commitments and presuppositions. Two familiar examples come to mind. First, our moral debates over the permissibility of withholding various life-sustaining treatments may well be shaped by assumptions premised on theories of causation. Baruch Brody has argued, for example, that our moral judgments in this area should largely track our metaphysical judgments about who causes what to happen *(52)*.

If it is morally wrong to cause someone's death in certain circumstances, we will need a theory of causation to help us sort out those instances in which human agents are responsible for deaths from those in which death might be causally attributed to other forces, such as the patient's underlying disease.

A second example of metaphysics in the service of bioethics comes from the theory of personal identity. Two potential applications come to mind. First, Michael Green and Dan Wikler have proposed an ingenious philosophical argument in favor of a so-called higher brain death criterion on the basis of their favored theory of personal identity *(53)*. The two leading candidates for the best account of personal identity, they argue, are (1) the "physical continuity view," which contends that our identities reside in the continuity of our self-same bodies over time, and (2) the "psychological continuity view," according to which our identity through time is established by the continuity of psychological states, such as memory. As partisans of the psychological continuity view, Wikler and Green regard the cerebral cortex as the reservoir of all the thoughts, memories, and other psychological states that provide the substrate for our identities as persons. Accordingly, in this view, once the cerebral cortex is destroyed, a person literally loses his or her identity as the person that he or she used to be. An individual may continue to subsist as a living human organism, but the *person* that he or she used to be is now dead. The most appropriate criterion for death, they conclude, is thus the destruction of the cerebral cortex or so-called "higher brain death."

The psychological continuity theory of personal identity has also been marshaled in the service of a moral theory of advance directives. According to John Robertson and Rebecca Dresser, the validity of advance directives depends *inter alia* on the continuity of personal identity between different stages of a person's life *(54)*. If the threads of memory holding our identities together are stretched or broken by dementia, stroke, or a coma, then an advance directive signed at time $T_1$ may no longer apply at a later time $T_2$ following a stroke. Indeed, the person at time $T_2$ may well be a *different* person than the person who signed the directive at $T_1$. Should this actually be the case, then following the advance directive will amount to not honoring the wishes of a self-same person, as the theory of advance directives would have it, but rather to the imposition of one person's wishes on another person! Dresser and Robertson conclude that in cases where we can no longer assume the continuity of personal identity, we should make decisions according to what is considered to be in the present patient's contemporaneous interests, not according to what the person would once have seen as being in *his or her* best interests.

How useful are such intriguing metaphysical theories for bioethics? Clearly, such theories of causation and personal identity can play a very large role indeed for academic bioethicists intent upon identifying the (often faulty) metaphysical presuppositions of our "commonsense" views on the nature of death or the validity of advance directives. It would, however, be extremely problematic to deploy these extremely abstruse, metaphysical debates in the service of clinical and policy-oriented bioethics. Although the prospect of controversy haunts the project of mid-level moral theory building in bioethics, that project at least attempts to touch base with widely shared moral intuitions about cases and principles. I doubt, however, that the same can be said for academic metaphysical disputes over the nature of causation and personal identity. Indeed, my hunch is that attempting to enshrine controversial metaphysical views in public policy would result in massive confusion and bad policy *(55)*.

## Metaethics and Bioethics

Another distinctly philosophical activity within bioethics falls under the heading of "metaethics." In contrast to those normative ethical theories that tell us in substantive terms what is good, bad, and virtuous in the domains of action and policy, metaethical inquiries have traditionally focused on both the meaning of various ethical terms—such as "good," "bad," "right," and "wrong"—and on the nature and methods of ethical inquiry itself. Although a great deal of contemporary metaethics is of scant interest to anyone with a practical bent, one variety of metaethical dispute has fuelled some of the most interesting controversies within bioethics during the past fifteen years. I refer here, of course, to our ongoing controversies over moral methodology. Despite the fact that a horde of methodological malcontents have attacked principlism for allegedly being "too philosophical," or for placing too much emphasis on principles and rules, the fact is that as long as these casuists, feminists, pragmatists, and narrativists recommend alternative ways of "doing ethics," they are all engaging in metaethics! Leading casuists, for example, argue that the real "locus of moral certainty" resides, not in mid-level principles, but rather in our ground-level moral intuitions of so-called paradigm cases *(56)*. They are, in other words, making claims about the nature of moral knowledge and whether it is general or particular. Likewise, some partisans of narrative ethics have claimed that the proper way to morally justify any proposed action is to place it in the context of our role-related duties, which in turn must be placed within the larger framework of the histori-cal narrative of our families, cities, states, and nations *(57)*. Thus, although the actual telling of such stories does not bear much resemblance to philosophical work, the claim that such stories provide a *better* grounding for our moral judgments than the prin-ciple of utility or categorical imperative is itself a distinctly philo-sophical claim.

Although these methodological debates are primarily of interest to specialists in the field rather than to your average mem-bers of a hospital ethics committee or bioethics commission, they do play an important role in shaping the nature of the field and of our self-understanding as bioethicists. Thus, if we accept the antiquated self-understanding of the field as a species of "applied ethics," we will see our job as involving the identification of the best ethical theory, followed by the deduction of conclusions by subsuming "the facts" under that theory. If, to the contrary, we embrace the rival account of justification known as reflective equilibrium, then we will be much more sensitive to the ways in which our judgments on the level of principle and theory can be modified by our responses to the particularities of moral contexts. If we accept the applied ethics model, then philosophers are likely to see themselves, if not as philosopher kings, then at least as the arbiters of what's kosher and what isn't in the field of bioethics. However, if we accept the model of reflective equilibrium, then philosophers are more likely to view themselves as the colleagues of social scientists, historians, physicians, lawyers, and medical humanists in a genuinely interdisciplinary enterprise *(58)*. Our methodological debates can thus have a profound impact on the kind of research that gets done in bioethics, and on how we view the nature of our collaboration with scholars from other disciplines.

## Conclusion

Does bioethics need philosophy? As I have labored to illustrate in this chapter, our response to this question depends on what kind of bioethical activity we have in mind (e.g., clinical, policy-oriented, or academic) *and* on how we understand the nature of philosophical theory. I have focused on two challenges to the proposition that bioethics needs philosophy: (1) the thesis that practical ethics—understood as a confluence of principlism, casuistry, and narrative—can and should remain independent of moral philosophy; and (2) the claim that standard brand, analytic philosophical ethical theory is incapable, beyond its narrow function of logical analysis, of guiding thought in practical contexts.

With regard to the first challenge, I have argued that bioethics is actually saturated with philosophical theory, and that a proper understanding of the methods of practical ethics will reveal all sorts of connections to higher level theorizing. With regard to the second challenge, I have largely accepted the criticisms of standard ethical theory while nevertheless insisting on the possibility and importance of other, less ambitious modes of philosophical theorizing for bioethics, including conceptual analysis, mid-level theory building, and metaethical reflection on methods. The conclusion that I draw from this exercise is that moral philosophy, shorn of its vaulting ambitions, can play a pivotal role in the construction of the interdisciplinary bioethical mosaic.

Before closing, I would like to suggest that the relationship between philosophy and bioethics is a two-way street. Not only does philosophy contribute to bioethics, but bioethics and practical ethics generally also have much to contribute to moral philosophy and ethical theory. This is especially true if one regards the latter activities as manifestations of *practical* reason—i.e., reason directed at the proper ends and means of *action*. Unfortunately, many who engage in normative ethical theory in the academy these days appear to have lost interest in the question of whether their theorizing bears any relationship whatsoever to actual human conduct. As a result, they never test their theories against concrete moral experience, so the theories remain purely abstract, academic (in the worst sense), and addressed to only a small handful of other like-minded and equally detached theorists.

Should we come to view the process of ethical justification as bearing more resemblance to the search for reflective equilibrium than to deductions from fundamental principles, the constructive role of practical ethics for philosophy comes immediately into view. With this approach to method, the task of justification involves the harmonization of all our moral commitments, ranging from our considered judgments of particulars to our theoretical constructions. Just as our more firmly held principles can force changes in our previously held intuitions about cases, so too our considered judgments about cases can force us to alter or abandon previously held theoretical commitments. Thus, from this methodological angle, the gap between practical ethics and ethical theory is effectively abolished, and practical ethics becomes both the origin and ultimate terminus of ethical theory.

Even if one wishes to remain agnostic on the merits of reflective equilibrium, ethical theorizing remains incomplete without moorings to practical moral experience. On the assumption that ethical theory is supposed to concern itself ultimately with the world of human action and social policy (i.e., assuming that it is not an end in itself) theory will remain blind to its own limits and shortcomings unless it draws from, and is tested against, practical problems *(59)*. Also, in the absence of serious engagement in the practical world, theorists will likely remain blind to pressing practical questions that can only be answered by serious and sustained theoretical work *(60)*. Neither result is salutary for the enterprise of philosophical ethical theory. Thus, not only does bioethics need philosophy, but philosophy needs practical ethics as well.

## Acknowledgment

I would like to thank Robert Crouch, Jennifer Flynn, and Jim Humber for their astute and most helpful comments on a previous draft of this chapter.

## Notes and References

[1]The language of law has had an equal, if not greater, impact on the field of bioethics. Indeed, I think it fair to say that philosophers have often played the role of conceptual custodians, sweeping out and tidying up the results of the day's court decisions.

[2]*See*, for e.g., Renée C. Fox and Judith P. Swazey. (1984) Medical Morality Is Not Bioethics: Medical Ethics in China and the United States. *Perspect. Biol. Med.* 27(3): 336–360; and John H. Evans. (2002) *Playing God: Human Genetic Engineering and the Rationalization of Public Bioethical Debate*. Chicago, University of Chicago Press, IL.

[3]Martha Nussbaum. Why practice needs ethical theory: particularism, principle, and bad behavior, in *Moral Particularism*. (Brad Hooker and Margaret Little, eds.) Oxford, Clarendon Press, pp. 233–234. Nussbaum's own definition of ethical theory does not insist that reasons and arguments be derived from a small number of fundamental principles. I have added this element to her account because I believe it captures an important feature of the kinds of traditional moral theories often invoked within bioethics.

[4]President's Commission for the Study of Ethical Problems in Medicine and Biomedical and Behavioral Research. (1983). *Deciding to Forego Life-Sustaining Treatment.*Washington, DC, US.

[5]*See*, e.g., Mark Kuczewski. (1997) Bioethics' consensus on method: who could ask for anything more? in *Stories and Their Limits*. (H. L. Nelson, ed.) Routledge, New York, NY.

[6]Alex London. (2001) The Independence of Practical Ethics. *Theoretical Medicine and Bioethics* 22(2): 87–105.

[7]K. Danner Clouser and Bernard Gert. (1990) A Critique of Principlism. *J. Med. Philos.* 15: 219–236.

[8]Albert Jonsen. (1991) Of Balloons and Bicycles, or The Relationship Between Ethical Theory and Practical Judgment. *Hastings Center Report* 21(5): 14–16. *See* generally Albert Jonsen and Stephen Toulmin. (1988) *The Abuse of Casuistry*. University of California Press, Berkeley.

[9]Cass R. Sunstein. (1996) *Legal Reasoning and Political Conflict*. Oxford University Press, New York, NY, 35 ff.

[10]Stephen Toulmin . (1981) The Tyranny of Principles. *Hastings Center Report* 11(6): 31–39.

[11]*See* the selection of articles in Hilda L. Nelson. *Stories and Their Limits*.

[12]Will Kymlicka. (1996) Moral philosophy and public policy, in *Philosophical Perspectives on Bioethics*. (L.W. Sumner and Joseph Boyle, eds.) University of Toronto Press, Toronto, pp. 244–270.

[13]*Ibid.*, p. 250.

[14]*Ibid.*, p. 252.

[15]*See* London. *The Independence of Practical Ethics.*

[16]I follow Rawls and Daniels in defining a "considered judgment" as a judgment made under conditions conducive to avoiding errors of judgment (e.g., while not under the sway of violent passion, on

the basis of adequate information, etc.). *See* Norman Daniels. (1979) Wide Reflective Equilibrium and Theory Acceptance in Ethics. *Journal of Philosophy* 76: p. 258.

[17]Derek Parfit. (1984) *Reasons and Persons*. Oxford University Press, Oxford.

[18]Ronald Dworkin. (1997) In Praise of Theory. *Arizona State Law Journal* 29: pp. 353–376.

[19]A good example of this sort of evasion can be found in National Institutes of Health, Report of the Human Embryo Research Panel (September, 1994).

[20]Ronald Dworkin. In Praise of Theory. p. 356.

[21]*Ibid.*, p. 358.

[22]*See*, e.g., Susan Wolf. (1995) *Feminism and Bioethics*. Oxford University Press, New York, NY; and Martha Nussbaum. (2000) *Women and Human Development*. Cambridge University Press, New York, NY.

[23]In this regard, *see* Susan Muller Okin's critique of Michael Walzer's communitarianism (1989), in *Justice, Gender and the Family*. Basic Books, New York NY, pp. 41–73.

[24]John Rawls. (1971) *A Theory of Justice*. Harvard University Press, Cambridge, pp. 20–22, 46–53.

[25]I shall bracket here the philosophical difficulties involved in equating coherence with justification. For an interesting and well-balanced discussion of this controversy, *see* Michael DePaul. (1993) *Balance and Refinement: Beyond Coherence Methods of Moral Inquiry*.Routledge, London/New York. *See also* Joseph Raz. (1994) The relevance of coherence, in *Joseph Raz, Ethics in the Public Domain*. Clarendon Press, Oxford, pp. 277–319.

[26]These various uses of reflective equilibrium—e.g., for justifying moral theories as well as practical judgments bearing on individual acts and policies—are well explored in Wibren Van Der Burg and Theo Van Willigenburg. (1998) *Relective Equilibrium*. Kluwer, Dordrecht.

[27]Tom Beauchamp and James Childress. (2001) *Principles of Biomedical Ethics, 5th ed.* Oxford University Press, New York, NY, pp. 397–401; and Dan Brock. (1996) Public moral discourse, in *Philosophical Perspectives on Bioethics*. (L. W. Sumner and Joseph Boyle, eds.) University of Toronto Press, Toronto, pp. 271–296.

[28]David DeGrazia. (1996) *Taking Animals Seriously: Mental Life and Moral Status.* Cambridge University Press, New York; and Baruch Brody. (1998) *Life and Death Decision Making.* Oxford University Press, New York, NY, Chapter 1.

[29]*See*, e.g., Tom Beauchamp. On Eliminating the Distinction Between Applied Ethics and Ethical Theory. *The Monist* 67(4): pp. 514–531; and Brock. Public Moral Discourse.

[30]For additional critiques in the same vein, *see* Robert L. Holmes. (1990) The Limited Relevance of Analytical Ethics to the Problems of Bioethics. *J. Med. Philos.* 15: 143–159: and Barry Hoffmaster. (1989) Philosophical ethics and practical ethics: never the twain shall meet, in *Clinical Ethics: Theory and Practice.* (Barry Hoffmaster, Benjamin Freedman, and Gwen Fraser, eds.) Humana Press,Totowa, NJ, pp. 201–230.

[31]Allen Buchanan, Dan Brock, Norman Daniels, and Daniel Wikler. (2000) *From Chance to Choice: Genetics and Justice.* Cambridge University Press, New York, NY, Chapter 5.

[32]Martha Nussbaum. *Why Practice Needs Ethical Theory: Particularism, Principle, and Bad Behavior.* p. 250. I give some concrete examples of such bad philosophy in the bioethics context below, at p. 25ff.

[33]Annette Baier. (1994)What do women want in a moral theory? in *Moral Prejudices.* Harvard University Press, Cambridge, pp. 2,3.

[34]Benjamin Freedman. (1987) Equipoise and the Ethics of Clinical Research. *New England Journal of Medicine* 317: pp. 141–145.

[35]National Bioethics Advisory Commission. (2001) *Ethical and Policy Issues in Research Involving Human Participants.* Bethesda, MD, Chapter 4.

[36]Later iterations of Freedman's project are shedding much needed light on current controversies over the ethics of international research trials. *See*, e.g., Alex John London. (2001) Equipoise and International Human–Subjects Research. *Bioethics* 15(4): pp. 312–332.

[37]Allen Buchanan and Dan Brock. (1986) Deciding for Others. *Milbank Quarterly* 64(2): pp. 67–80.

[38]Buchanan and Brock argue that competency determinations are "all or nothing" in the sense that they either allow or disallow a patient to make a given decision.

[39]Bruce Miller. (1981)Autonomy and the Refusal of Lifesaving Treatment. *Hastings Center Report* (11)4: pp. 22–28.

[40]For the sake of brevity, I shall only mention in passing here a fourth example of important conceptual analysis for bioethics—namely, the analysis of such ubiquitous, yet little understood, concepts as "coercion" and "exploitation." Especially in the contexts of reproductive and research ethics, these concepts have lately been thoroughly debased to the status of all-purpose negative epithets. For two extremely helpful conceptual–ethical remedies, *see* Alan Wertheimer. (1987) *Coercion.* Princeton University Press, Princeton; and Alan Wertheimer. (1996) *Exploitation.* Princeton University Press, Princeton.

[41]Norman Daniels. (1985) *Just Health Care.* Cambridge University Press, New York.

[42]*Ibid.*, p. 41.

[43]Nor need it be, as some uncharitable critics have alleged, a fraudulent attempt to portray one's personal political preferences as the conclusions of a "scientifically neutral" moral philosophy. *See*, e.g., Holmes. (1985) The Limited Relevance of Analytical Ethics To the Problems of Bioethics; and Annette Baier. (1985) Doing Without Moral Theory? *Postures of the Mind.* University of Minnesota Press, Minneapolis, pp. 228–245.

[44]Norman Daniels and James Sabin. (1997) Limits to Health Care: Fair Procedures, Democratic Deliberation, and the Legitimacy Problem for Insurers. *Philosophy and Public Affairs* 26(4): 303–350. For a more generalized vindication of the claim that ethics and political philosophy must eventually be supplemented by democratic deliberation, *see* Amy Gutmann and Dennis Thompson. (1996) *Democracy and Disagreement.* Harvard University Press, Cambridge.

[45]Joel Feinberg. (1984) *Harm to Others.* Oxford University Press, New York.

[46]Tom Beauchamp and Ruth Faden. (1986) *A History and Theory of Informed Consent.* Oxford University Press, New York, NY.

[47]Beauchamp and Childress. *Principles of Biomedical Ethics, 5th ed.* pp. 293–311.

[48]Gerald Dworkin. (1988) *The Theory and Practice of Autonomy.* Cambridge University Press, New York, NY.

[49]Allen Buchanan and Dan Brock *Deciding for Others.*

[50]Some academic bioethicists have much more faith in the dictates of their unconstrained theories than they do in regnant medical and social norms. Thus, although their theorizing is important and interesting as an academic exercise, it largely fails to connect with the social and political world as we know it, and thereby forfeits the ability to influence policy. Good examples of this kind of unconstrained academic theorizing can be found in the works of Peter Singer, an unconstrained utilitarian, and H. Tristram Engelhardt, Jr., an unconstrained libertarian.

[51]For a similar defense of middle-level theory, *see* Ruth Macklin. (1989) Ethical theory and applied ethics: a reply to the skeptics, in *Clinical Ethics: Theory and Practice.* (Barry Hoffmaster, Benjamin Freedman, and Gwen Fraser, eds.) Humana Press, Totowa, NJ, pp. 101–124.

[52]Baruch Brody. *Life and Death Decision Making.* pp. 25,113–115.

[53]Michael Green and Daniel Wikler. (1980) Brain Death and Personal Identity. *Philosophy and Public Affairs* 9(2): 105–133.

[54]Rebecca Dresser and John Robertson. (1989) Quality of Life and Non-Treatment Decisions for Incompetent Patients. *Law, Medicine and Health Care.* 17(3): 234–244.

[55]*See,* e.g., Allen Buchanan. (1988) Advance Directives and the Personal Identity Problem. *Philosophy and Public Affairs* 17: 277–302.

[56]*See* Jonsen and Toulmin. *The Abuse of Casuistry.* p. 330.

[57]*See*, e.g., David Burrell and Stanley Hauerwas. (1977) From system to story: an alternative pattern for rationality in ethics, in *Knowledge, Value and Belief.* (H.T. Engelhardt, Jr., and Daniel Callahan, eds.) The Hastings Center, Hastings-on-Hudson, NY.

[58]For some ruminations on Dewey's potential contribution to this debate over the proper role of the philosopher in practical ethics, *see* John D. Arras. (2002) Pragmatism in Bioethics: Been There, Done That. *Social Philosophy and Policy* pp. 39–41.

[59]A good example of this problem is the hard won insight that standard theories of justice are ill equipped to help us answer the most basic problems of health care rationing. *See* Daniels and Sabin. *Limits to Health Care: Fair Procedures, Democratic Deliberation, and the Legitimacy Problem for Insurers.*

[60]An example of this problem is the failure of political theorists to recognize an issue raised sharply by advocates for the so-called disabilities rights movement—namely, how inclusive of differences
should a society's fundamental institutions be? *See* Allen
Buchanan et al. *From chance to choice*, in *Genetic Intervention
and the Morality of Inclusion.* Chapter 7.

# 2

# Religion, Theology, and Bioethics

*James F. Childress*

## Myths of the Role of Religion in the Origin of Bioethics

This chapter focuses on the roles and potential contributions of religion, theology, and religious studies in their related but distinctive ways to bioethics *(1)*. Medical ethics often refers to ethics for physicians, analogous to ethics for other health care professionals, such as nurses. Over time, various professional groups in health care have codified ethical standards to guide practice. Many religious communities have also provided ethical guidance for their members as caregivers and as patients. Indeed, the lines have not always been clearly drawn between these sources of guidance. For instance, the original Hippocratic oath was probably articulated by a Pythagorean cult *(2)*. Beyond ethics for health care professionals (and, sometimes, for patients and families), "bioethics" or "biomedical ethics" emerged in the late 1960s and early 1970s to address the moral perplexities engendered by new medical technologies that could prolong life far

From: *The Nature and Prospect of Bioethics: Interdisciplinary Perspectives*
Edited by: F. G. Miller, J. C. Fletcher, and J. M. Humber
© Humana Press Inc., Totowa, NJ

beyond previous expectations; transplant organs from one person to another; detect certain fetal problems in utero; offer new reproductive possibilities, and the like. Bioethics generally examines the ethics of acts and practices in the life sciences, medicine, and health care. Despite a general distinction between medical ethics and bioethics, the terms are frequently used interchangeably. However the lines are drawn, the intellectual activity of bioethics is not limited to professional bioethicists, whether theologians, or philosophers, or others. It is multidisciplinary and multiprofessional.

Theologians and philosophers, along with various other professionals, have contributed significantly to the development and evolution of bioethics. Nevertheless, various stories of the origins of bioethics provide different interpretations of the role(s) of religion and theology in its emergence and evolution. One myth of origins highlights religion, as articulated particularly by several founders of the field who were theologians, or at least religiously oriented philosophers, scientists, and clinicians. For example, regarding the respective contributions of philosophers and theologians, Albert Jonsen observes that "Theologians were the first to appear on the scene" *(3)*.

This first myth of origins claims that theologians, or religiously oriented philosophers and other religiously oriented professionals, contributed to a "renaissance of medical ethics" and to the creation of bioethics *(4)*. Examples of this myth are numerous. One of the founding figures in bioethics, Daniel Callahan, a philosopher with religious interests, writes: "When I first became interested in bioethics in the mid-1960s, the only resources were theological or those drawn from within the traditions of medicine, themselves heavily shaped by religion" *(5)*. Key figures included James Gustafson, Paul Ramsey, Richard McCormick, S.J., Immanuel Jakobovits, and William May, among others.

Proponents of this first myth of origins usually note the subsequent "secularization of bioethics" or "marginalization" of religious and theological voices in bioethics. Several critics of contemporary bioethics lament this development. Drawing on Daniel Callahan, two theological critics of the secularization of bioethics trace the decline of religion's and theology's role in the development of bioethics:

> ...after the "renaissance of medical ethics" came the "enlightenment" of medical ethics. In the next decade, interest in religious traditions moved from the center to the margins of scholarly attention in medical ethics. The theologians who continued to contribute to the field seldom made an explicit appeal to their theological convictions or to their religious traditions. *(6)*

The putative causal factors include the difficulty of doing bioethics in a pluralistic society, whether at the bedside or in the public square, and as a consequence, the growing emphasis on principles, rights, and procedures.

Recently, Carla M. Messikomer, Renee C. Fox, and Judith P. Swazey have challenged the "'religious-to-secular' trajectory of bioethics" as "not quite accurate" and also "over-simplified in a number of ways" *(7)*. Indeed, they seek to dispel the conventional myth about the historical role of religion and theology in the origin of bioethics, claiming that it exaggerates their role. To be sure, they recognize religious contributions from theologians and religious ethicists, as well as from other professionals including philosophers, such as Hans Jonas, and clinicians. However, they contend that religion's shaping influence on the conceptual framework of bioethics, its issues, its substance and form, and its relationship to public-policy deliberations was "modest at best." Indeed, numerous religious ethicists and theologians were "already conforming to what quickly became [the field's] predominant, rational secular mode of thought." The overall conceptual framework was largely secular with an orientation toward individual rights, in part because American bioethics emerged in the era of the civil rights and anti-war movements, and because of the felt need at the time to find consensus in the public square through principles and other measures. According to interpreters, the field of bioethics has had a strong and persistent proclivity to "screen out" religious questions, or "to 'ethicize' them by reducing them to the field's circumscribed definition of ethics and ethical."

Messikomer and colleagues reach a different conclusion to the standard myth, in part because they have a different conception of religion. Their conception does not focus on "religious doctrine, practice, or membership in a particular faith community." It is not tradition based. Instead, it attends to "basic and transcendent aspects of the human condition, and enduring problems of meaning to questions about human origins, identity, and destiny; the "why's" of pain and suffering, injustice and evil; the mysteries of life and death; and the wonders and enigmas of hope and endurance, compassion and caring, forgiveness and love." Furthermore, various "metathemes," particularly concerning human birth and mortality, personhood, and finitude, are "religiously resonant, independently of how they are viewed by the articulators and practitioners of bioethics."

In short, these authors claim, apart from a few bioethicists, that religion in the broad, formal sense has never really been very significant in bioethics which, throughout its first three decades, has "characteristically 'ethicized' and secularized, rationalized and marginalized religion, and thereby restricted its influence on the mode of reflection and discourse, and the purview of the field." Messikomer and colleagues wonder whether more recent developments, such as the increasing attention to religion in medicine, might augur some shifts in the patterned ways bioethics has accommodated and domesticated religion.

This conception of religion is problematic, in part because these "aspects," "problems," and "questions," are construed as religious, whatever interpretations, solutions, or answers are provided by those who address them. This 'formal' definition of religion obviously encompasses a broader span of human activities than a substantive definition would—for instance, a substantive definition that focuses on sacred or transcendent powers. However, somewhat paradoxically, in the story told by Messikomer and colleagues, the broad, formal definition allows them to exclude some positions that adherents deem to be religious, and that relate bioethics to sacred or transcendent beliefs, norms, and rituals, as interpreted in particular traditions. More importantly, it allows them to maintain that religion has played a modest role throughout the history of bioethics because of bioethics' limited attention to certain problems and questions.

Although quite provocative, this story suffers from an indefensibly broad, formal definition of religion that makes any and all positions "religious," even if they claim to be secular, because of the problems or questions they address. It similarly excludes positions that their adherents view as religious *(8)*. Their argument would have been more persuasive if they had avoided the religious–secular polarity altogether and argued instead that bioethics has not adequately addressed certain problems and questions, whatever label is used.

Those who lament the limited role of religion and theology in American bioethics, whichever myth they defend, often charge that this limited role results, in part, from the appeal in bioethics to broad ethical principles and individual rights. For example, some contend that forms of principlism, with intermediate or mid-level principles, such as respect for autonomy, beneficence, non-maleficence, and justice, impede religious and theological contributions to bioethics. However, at least as interpreted by Tom Beauchamp and me in *Principles of Biomedical Ethics*, such principles are arguably justifiable from the standpoint of different religious and secular traditions, or represent an overlap or convergence of those traditions, and particular traditions fill out these principles in their thicker and richer interpretations of persons, harm, benefits, etc. *(9)*.

It is plausible to argue that religious and theological discourse did become marginalized within the field of bioethics as a whole before being at least partially restored. However, even this argument requires a careful and nuanced statement. In particular, religious communities, theologians, rabbis, and other interpreters of their traditions continued to address bioethics in light of their accepted sources of authority, in order to guide professional and non-professional caregivers and patients, as well as—in some cases—to shape the broader culture and to influence public policy. Bioethics is not monolithic and, certainly, much hinges on which segment of bioethics is taken to represent the field as a whole.

Nevertheless, there are signs of growing attention to religion and theology in bioethics in general, and that increased attention, within limits, is warranted for several reasons. I explore these reasons in three contexts: (1) interactions between professionals and patients, clients, and others; (2) culture; and (3) public policy. These are not mutually exclusive contexts, but they do allow us to identify several ways in which religion, religious studies, and theology can contribute to bioethics. Indeed, bioethics as a whole needs to attend more to religious and theological perspectives in all three contexts.

## Religion in Professional Caregiver and Patient Interactions

### *Patients*

The principle of respect for persons, including their autonomous choices, is widely accepted in bioethics in the United States— some commentators even claim that it has become the dominant principle. However, most major bioethical theories recognize the need to consider other weighty principles as well. It is not possible to apply this principle mechanically because people are complex, and discerning their preferences often requires a difficult interpretive activity, in part because persons participate in various communities whose beliefs, norms, and practices they accept wholly or in part. Some of those communities are religious, and attention to the particularities of religious traditions can aid clinicians' efforts to respect, as well as to benefit, their patients.

For example, questions arise as to when, if ever, it is justifiable for clinicians to override, or to seek a court order to override, the religiously based health care decisions of Jehovah's Witnesses, Christian Scientists, faith healers, and others. While these questions may arise with respect to the care of adults, especially with dependents, such situations are less problematic now. However, parental or surrogate decisions for children and other persons with limited decision-making capacity continue to pose vexing questions *(10)*. The need for greater familiarity with particular religious traditions becomes evident when some philosophers confuse, for example, Jehovah's Witnesses and Christian Scientists, whose fundamental beliefs and practices differ so much *(11)*. In general, Jehovah's Witnesses accept most medical treatments, drawing the line at blood transfusions, while Christian Scientists refuse a wide range of medical treatments. Not only do their beliefs and practices differ, but their religious–theological frameworks are also fundamentally different. Furthermore, they pose different practical, ethical problems because Jehovah's Witnesses typically enter the system of medical care and then set limits on what may or may not be done, while Christian Scientists tend to avoid the system of medical care altogether. Thus, greater familiarity with such religious traditions would enable clinicians and bioethicists to better interpret patient preferences and determine when, if ever, to override those preferences, especially to protect children.

However, as Dena Davis reminds us, it is not sufficient only to study religious traditions' official beliefs, norms, and practices, because some individuals who identify with those traditions have quite different conceptions *(12)*. As I often put it, even those whose religious views are considered heretical within a particular tradition have a right to respect in the context of medical care. To take one example, according to traditional Navajo beliefs, negative information, such as the disclosure that a procedure could have serious side effects, may actually cause the problems. After all, within this framework, language does not merely reflect reality; it creates reality *(13)*. However, caution is needed in moving from a general interpretation of traditional Navajo beliefs and practices to a judgment about what a particular Navajo patient would want. For instance, it would be a mistake for a physician in the Indian Health Service simply to withhold or limit information about the risks of a medical procedure from a particular Navajo patient on such grounds—that particular patient may have rejected the traditional position.

Thus, acting on a principle of respect for persons, including their autonomous choices, requires attention to religious beliefs, norms, and practices—in contrast to those who would argue that principles limit attention to religion. Furthermore, religious studies, particularly comparative religious studies, can greatly enrich bioethical interpretation and guidance.

## *Clinicians*

Clinicians may also need and want to understand how their own religious traditions—beliefs, practices, and norms—bear on their own actions. For example, does a clinician's religious tradition permit participation in abortion and, if so, under what conditions? Within a particular religious tradition, would certain research on embryonic stem cells constitute complicity with wrongful practices of abortion or wrongful destruction of embryos following in vitro fertilization? When would conscientious refusal or even civil disobedience appear to be justified, or even mandated by the tradition? Such reflections have long been part of traditional medical ethics, as articulated within particular religious communities.

For bioethicists, familiarity with religious studies can be helpful for providing advice and counsel for clinicians. Such studies can provide useful insights into the beliefs, practices, and norms of particular religious traditions. And they should also include attention to the theologies that are articulated by thinkers within those traditions. Through such studies, bioethicists can become better conversation partners with clinicians who are seeking to understand their own traditions, and can also make a significant contribution to clinicians' religious–moral reflection and deliberation by directing attention to the best available resources.

## Religion, Theology, and Culture

When presenting their arguments about particular positions, bioethicists generally seek to have an impact on culture. According to a classic definition, "Culture consists of patterns, explicit and implicit, of and for behavior acquired and transmitted by symbols, constituting the distinctive achievement of human groups" *(14)*. In many respects, in the so-called "culture wars," bioethical issues have become a primary battleground for conflicts. Commentators often present these conflicts in terms of opposition between religious and secular viewpoints in bioethics, but the conflicts appear within religious traditions as well, as has been evident in debates about human embryonic stem-cell research. Indeed, some sociologists suggest that liberals (or conservatives) in Protestantism, Roman Catholicism, and Judaism may share more with liberals (or conservatives) in the other traditions than with their own religious colleagues who do not share their liberal (or conservative) orientation *(15)*. Hence, despite some media interpretations, the so-called "culture wars" are not waged by "enemies" who can be easily identified by specific religious labels.

When theologians consider bioethical issues, they use a variety of approaches that address culture in different ways. Although each of these approaches has distinctive features, they are not all mutually exclusive. I draw a few significant examples from Christian theological reflection, while noting that there are parallels in Jewish and other traditions too, and stressing that comparative bioethics offers one of the most promising paths of exploration over the next several years. My examples include theological bioethicists who view their task as more than being "moral philosophers with a special interest in 'religious texts and arguments'"*(16)*.

The late Paul Ramsey was one of the most important Protestant voices in medical ethics, research ethics, and, more generally, bioethics. He waged legendary battles with other ethicists, including Joseph Fletcher, who over time moved progressively toward a non-religious ethic, stated largely in utilitarian terms, and several of those battles focused on bioethics. Ramsey did not develop the theological foundations for his bioethics as thoroughly as many thought he should have; some even charge that most of the theology in his influential book, *The Patient as Person*, appears in its brief preface, which sketches his covenantal perspective *(17)*.

In this book and elsewhere, Ramsey recognized several intermediate principles or rules that could provide "ethical bridge-work" between theological ultimates and practical decisions; this "ethical bridgework" largely consisted of deontological constraints. Attending less to ends and consequences, Ramsey brought various deontological norms under covenant responsibilities. Speaking specifically about the practice of medicine as a covenant, he noted:

> Justice, fairness, righteousness, faithfulness, canons of loyalty, the sanctity of life, hesed, agape or charity are some of the names given to the moral quality of attitude and of action owed to all men by any man who steps into a covenant with another man—by any man who, so far as he is a religious man, explicitly acknowledges that we are a covenant people on a common pilgrimage. *(18)*

Ramsey's approach to medical ethics starts from the social institutionalization of religious beliefs, norms, and practices. We now encounter historical deposits or embodiments of religious–

ethical norms in the broader society and culture. And now believers and non-believers alike can appeal to those norms. For instance, Ramsey writes:

> ...the Judeo-Christian tradition decisively influenced the origin and shape of medical ethics down to our own times. Unless an author absurdly proposes an entirely new ethics, he is bound to use ethical principles derived from our past religious culture. In short, medical ethics nearly to date is a concrete case of Christian "casuistry"—that is, it consists of the outlooks of the predominant Western religion *brought down to cases* and used to determine their resolution" (italics in original) *(19)*.

Stanley Hauerwas interprets Ramsey's position in an overstatement that nevertheless makes his point effectively:

> Medicine, at least his [Ramsey's] account of medicine, confirmed his presumption that agape was in fact instantiated in Western culture. In effect, medicine became Ramsey's church as doctors in their commitment to patients remained more faithful to the ethic of Jesus than Christians who were constantly tempted to utopian dreams fueled by utilitarian presumptions. *(20)*

In approaching these historical deposits, Ramsey sought largely to interpret, extend, deepen, and refine them, rarely to reject or fundamentally revise them. Thus, among other topics, he presented arguments about standards for withholding and withdrawing life-prolonging treatment and practices of obtaining organs for transplantation that could be debated without appealing to fundamental theological commitments, however important those commitments may have been for the positions he took.

Whereas Ramsey tended to focus on moral quandaries, undertaking casuistical analysis in light of historically embedded religious–moral norms, Protestant William F. May construes the task of ethics as one of "corrective vision." He often focuses on beliefs, symbols, and rituals—some explicitly religious and others implicitly religious—offering a theological critique wherever warranted. For example, in a debate with philosopher Joel Feinberg about practices of obtaining organs for transplantation—with May arguing for express consent and Feinberg proposing routine salvaging—May takes seriously emotional responses expressed in symbols and rituals *(21)*. Feinberg charges that May's approach resembles that of "literary critics debating the appropriateness of symbols" even as people are dying *(22)*. Feinberg calls for a "rational superintendency" of emotions and sentiments in order to attend to the needs of people who are dying because of a lack of transplantable organs. May responds that rationality alone is inadequate, that we lack a symbol-free access to reality, that symbols have "cognitive and moral significance," and that symbols and rituals enable us to direct and discipline our sentiments, not merely to express them.

According to Gilbert Meilaender's interpretation, May has adapted:

> ...a style that makes possible a theological ethic that addresses fundamental questions without denying whatever human insight can be found. He draws on as many sources of wisdom as he can—while never failing to provide the corrective re-envisioning that his theology makes possible and that, he judges, it is the task of ethics to provide. ...To him there is nothing human that is not in need of correction and transformation when related to the transcendence of God. *(23)*

Meilaender, a Lutheran theological ethicist, works more directly from the stories and concepts of the Christian tradition. Speaking in a public-policy context, in testimony to the National Bioethics Advisory Commission (NBAC) about cloning, but with particular attention to public culture, he underscores how reflection within a particular tradition can disclose the universal:

I will speak theologically—not just in the standard language of bioethics or public policy. I do not think of this, however, simply as an opportunity for the "Protestant interest group" to weigh in at your [NBAC's] deliberations. On the contrary, this theological language has sought to uncover what is universal and human. It begins epistemologically from a particular place, but it opens up ontologically a vision of the human. The unease about human cloning that I will express is widely shared. I aim to get at some of the theological underpinnings of that unease in a language that may seem unfamiliar or even unwelcome, but it is language that is grounded in important Christian affirmations that seek to understand the child as our equal—one who is a gift and not a product. ...I will also suppose that a faith [that] seeks understanding may sometimes find it. *(24)*

The relation of reason and faith is also central in Roman Catholic moral thought. As interpreted by Richard McCormick, S.J., the natural law provides the general, essential norms that apply to all human beings. But faith informs reason, without abolishing it, and theology informs bioethics, largely through Christian narratives, which include claims about God as the author and source of life, human destiny, Jesus' life, death, and resurrection, etc. The new light that faith casts directs us to solutions in bioethics and elsewhere that are "fully human." In this connection, theology can help in several ways by, for example, directing us to view life as a basic but non-absolute good *(25)*.

Eschewing claims about an objective and universal natural law, some religiously oriented bioethicists narrow their attention to a particular religious community and seek to develop the implications of its beliefs, norms, and practices for its adherents in facing bioethical issues. Two prominent examples are Stanley Hauerwas (a Methodist who is indebted to various other strands of the Christian tradition) and H. Tristram Engelhardt, Jr. (an Orthodox Christian), both of whom understand their respective versions of the Christian tradition to be countercultural *(26)*.

In short, there are many different kinds of religiously and theologically oriented bioethics, not all of which are subject to the same criticisms. For example, R.C. Lewontin, a scientist, praises philosophical reflections in bioethics while denigrating religious reflections. He claims that theologians attempt to "abolish hard ethical problems" and avoid "painful tensions" *(27).*[*] Many theologians, as well as many philosophers, recognize "painful tensions." Neither group as a whole fails to appreciate the moral conflicts involved in, for example, human reproductive cloning in various scenarios, or in different public policies toward human reproductive cloning, even if different thinkers resolve them differently. Indeed, in contrast to Lewontin's claim, religious and theological ethicists frequently complicate the moral picture, rather than simplify it. Instead of seeking philosophical clarity, they often seek illumination through religious stories, concepts, symbols, and rituals, among other aspects of religion, which resist tidy philosophical analysis.

I, for one, do not believe that clarification and illumination are opposed. Thus, I want to bring philosophical bioethics and religious/theological bioethics into closer interaction and dialogue. Religious/theological bioethics can benefit greatly from the best available philosophical analysis and argumentation, while philosophical bioethics can benefit greatly from religious/ theological perspectives.[**] Sometimes explicit theological reflection can generate moral insights that non-religious persons can endorse without accepting specific religious–theological starting points and premises. Engagement of non-religious bioethics with this kind of approach can be quite illuminating and

---

[*]I will ignore his phrase "internal contradictions," which he includes along with "painful tensions," because both philosophers and theologians try to avoid "internal contradictions" in order to develop defensible positions.

[**]My own work concentrates on the former, whereas this essay mainly addresses the latter because of its assigned topic.

should not be avoided merely because they are religious–theological in nature. Clearly the kind of engagement that is possible and fruitful will vary greatly, depending in part, for example, on where the religious–theological position falls in the range of positions I briefly sketched.

One reason sometimes given for why we need more attention to religious/theological bioethics is that secular or humanistic bioethics, informed by philosophy, tends to accept the status quo, rather than to challenge it. However, it is not clear that different degrees of acceptance of or resistance to scientific and technological developments correlate with religious/theological and secular/philosophical perspectives. Nevertheless, the prophetic voice is one voice that traditions such as Judaism, Christianity, and Islam recognize and affirm in different ways.

The prophetic voice is one of four voices in bioethics identified by James M. Gustafson *(28)*. One is narrative: this captures part of Hauerwas' work, but it could also extend to some forms of casuistry. A second is ethical analysis. Much of Ramsey's and McCormick's work in bioethics took this form. A third voice is policy, which we will return to in the last section of this chapter. The fourth voice is prophetic: this last voice is most explicitly counter-cultural. Much of Engelhardt's religiously oriented bioethics (in contrast to his libertarian approach to bioethics in the modern state) and Hauerwas' theology is prophetic in nature. Other prophetic voices also criticize our sociocultural responses to biomedical technology, particularly our glorification of that technology as part of the modern project *(29)*. These and other critics charge that much of bioethics, in part because of the decline of religious perspectives, has become priestly (blessing technological developments) and regulatory (seeking to regulate those developments). Certainly attention to public policy, and not only to culture, has characterized much bioethics, whether philosophical or theological in nature, and serious questions arise as to the appropriate role of religion and theology in this context.

# Religious Convictions in Public Policy
# in a Liberal, Pluralistic Society

What is the proper role for religious convictions in the formulation of bioethical public policy in a liberal, pluralistic democracy? This is a question in applied or practical political philosophy, rather than in theology, even though theologians also address it. This question concerns the value and limits of attending to religious perspectives, or listening to religious voices, in formulating or recommending public policies in a liberal, pluralistic, democratic society. We can address this question most fruitfully by focusing on public justification within our own liberal, pluralistic democracy with its particular history, traditions, etc., rather than trying to provide an answer for all conceivable democracies.

Within our own democratic polity, this question can arise in at least two ways. On the one hand, it may arise when adherents to a particular religious tradition claim exemption from some legally mandated conduct, such as providing conventional medical care for a child. For example, it emerges when Jehovah's Witness parents refuse a blood transfusion on behalf of their child even though, according to the best medical advice, that transfusion is necessary to save their child's life. To take another example, it emerges when Native Americans and Orthodox Jews, among others, object to the application of the legal standard of "brain death" *(30)*.

On the other hand, this question may also arise when participants in a particular religious community seek to shape the direction and content of bioethical public policy for everyone, rather than merely seeking an exemption for themselves from some legally mandated conduct. I want to outline a perspective on this second question. I have developed this perspective more fully elsewhere, in part in reflection on the decision by the National Bioethics Advisory Commission (NBAC), on which I served as a member, to ask scholars within particular religious traditions to present the views of those traditions on human reproductive cloning and on human embryonic stem-cell research—two scientific and technological breakthroughs that President Clinton specifically asked NBAC to address in reports and recommendations for federal policy *(30)*.

In discussing the prospect of human cloning, President Clinton commented that "any discovery that touches upon human creation is not simply a matter of scientific inquiry, it is a matter of *morality* and *spirituality* as well." Shortly thereafter, NBAC set up two days of hearings, with particular attention to religious, as well as other ethical and scientific perspectives. NBAC recognized that public policy in the United States cannot be based on purely religious considerations—that its own reasons for its recommendations could not, and should not, be religious. However, for several reasons, commissioners believed that it was important to consider religious perspectives, along with philosophical perspectives, on human cloning, as it analyzed and assessed various moral arguments and deliberated about public policies.

In two places in its report on cloning humans, NBAC listed several reasons, in slightly different language, for attending to religious perspectives: religious traditions "influence and shape the moral views of many US citizens, and religious teachings over the centuries have provided an important source of ideas and inspirations"; "policy makers should understand and show respect for diverse moral ideas regarding the acceptability of cloning human beings in this new manner"; often religious ideas "can be stated forcefully in terms understandable and persuasive to all persons, irrespective of specific religious beliefs"; NBAC wanted to determine whether "various religious traditions, despite their distinctive sources of authority and argumentation, reach similar conclusions about this type of human cloning" because a convergence among them, as well as among secular traditions, would be instructive for public policy; "all voices should be welcome to the conversation"; and a range of moral views need to be considered in determining the feasibility and costs of different policies which might be affected, for example, by vigorous moral opposition *(32)*.

These unsystematic reasons implicitly challenge some conceptions of "public reason" that seem, or even seek, to exclude religious convictions from public decisions and justifications, and they could and should be developed more systematically. Without attempting to undertake this systematic task at this juncture, I want to identify a few additional reasons that merit attention.

First, public reasoning includes imagination, not only rational deduction from shared secular premises. Religious stories and theological concepts may enable us to imagine and re-imagine in ways that are fruitful to public policy. For example, in testifying before NBAC, Christian and Jewish thinkers explored the meaning and significance of human reproductive cloning, as well as the moral status of the unimplanted, early embryo, and they did so in part by appealing to religious stories and concepts—the stories in Genesis were especially important, even though speakers emphasized different stories and offered different interpretations of these stories. Overall their testimony provided rich perspectives on the technology for cloning human.

Hence, imagination may be important for public policy, even if in the end policy-makers or advisors reject the positions that the religious stories and concepts support, and appeal only to secular grounds. It may also occur even when the religious positions diverge, as they did in many respects in testimony before NBAC, particularly on human embryonic stem-cell research. And it may occur even if we seek secular equivalents or translation into secular language. However, a strictly rationalistic model of public reasoning tends to neglect the important, if complex, role of imagination in public policy.

Second, religious communities and traditions often present moral positions and arguments that are not merely, or exclusively, religious in nature—this point was evident in several of the approaches to bioethics I sketched in the previous section. For instance, a judgment about human cloning may appeal to "nature" or "basic human values" or "family values," all categories that are not reducible to particular faith commitments, and that may be accessible to citizens of different or no faith commitments. Indeed, religious and nonreligious communities and individuals may share certain norms in an overlapping consensus, even though they would justify those norms in quite different ways. In addition, a widespread, overlapping consensus appears in the conclusion many have reached that it is unethical, at least for now, to engage in somatic-cell nuclear transfer cloning to create children.

Third, even the most vehement critics of appeals to religious convictions in public reasoning about bioethical policies, particularly policies that potentially involve coercion, generally concede a legitimate place for religious convictions in the "background culture," which according to John Rawls includes "non-public reasons...the many reasons of civil society" that obviously shape how a new technology such as human reproductive cloning is viewed *(33)*. However, it may be very difficult to draw a hard and fast line between "background culture" and public political debate, particularly on new technologies such as human reproductive cloning. Indeed, the "background culture" tends to spill over into public debate about appropriate policies; hence, it may be better to welcome all arguments and assess their worth in public.

One version of this last argument also appeals to specific features of the background culture and political life in the United States. Michael Perry notes that many citizens in the United States are religious, according to various surveys, and stresses that it is unreasonable to suppose that their religious convictions will not figure into their judgments about public policies. He thus makes the following argument: "Because of the role that religiously based moral arguments inevitably play in the political process then, it is important that such arguments, no less than secular moral arguments, be presented in, so that they can be tested in public, political debate" *(34)*. Otherwise citizens will set aside, in public discourse, the real reasons for their positions, and it will be impossible to assess and test those hidden reasons.

When religious convictions enter the public square, they are subject—and should be subjected to—close public scrutiny, just as any other reasons. For instance, some of them may be mutually contradictory. The fact that a position is religiously based gives it no special claim or privilege in public, political debate. Neither exclusion nor privilege is appropriate. A model of "religious and moral engagement" (in Michael Sandel's language) which is defensible as a form of "mutual respect" in a deliberative democracy, entails both inclusion and scrutiny *(35)*. However, it is not easy to determine the appropriate tests of religious–moral positions, in part because some of their premises may not be susceptible to rational adjudication.

In my judgment, it is consistent with a strong conception of liberal democracy to attend to religious perspectives and voices in the *process* of formulating public policies, as NBAC attempted to do in an advisory capacity, but it is also necessary to limit their role in the *content and justification* of those policies—again, as NBAC also attempted to do. The *process* of reaching a decision— or in NBAC's case, a recommendation—should consider the widest possible range of positions and reasons for those positions, but the *outcome* in substance and public justification needs to involve a *sufficient* or *adequate* secular, i.e., non-religious reason *(36)*. This is especially true when the policy involves coercion, as so many do—for example, a proposed criminal ban on human reproductive cloning.

While my various points fit well with NBAC's approach to bioethical public policy, some critics charge that both are reductionist and anticipate that other commissions might, in some unspecified way, better include religion/theology. For example, some critics contend that, in the final analysis of NBAC's report on human stem-cell research, religion was "expunged by being reduced to 'diverse perspectives,' 'ethical issues,' and 'moral concerns'" *(37)*. Still others believe that a return to substantive rationality, focused on ends and conceptions of the good, would both open the door to and benefit from religion and theology. They stress the value of thicker discussions available in religion and theology *(38)*.

At the time of this writing (May 2002), it is too early to tell how the President's Council on Bioethics (PCB), established by President Bush to succeed NBAC, will handle religious convictions and theology in its reports. The PCB's discussion at its first three meetings, which focused to a great extent on human reproductive cloning and human embryonic stem cell-research has been broad, and religious convictions and theological perspectives have been evident. The PCB includes theologians, such as William F. May and Gilbert Meilaender, as well as other members who are explicitly and closely tied to particular religious traditions and positions. However, if the PCB recommends public policies, it is unclear whether, or how, it will incorporate religious views. If a bioethics commission attempts to formulate and justify a policy about the use of federal funds or about coercive laws on the basis of particular religious convictions, even with a broad conception of religion, the content and justification are problematic and perhaps unfair and disrespectful of fellow citizens in our liberal, pluralistic democracy.

## Conclusion

I began by identifying two competing myths of the role of religion and theology in the origin of bioethics in the late 1960s and early 1970s, noting that contested definitions of religion and different conceptions of bioethics appear in these different myths. Then I argued that increased attention to religion in bioethics, through both religious studies and theology, would be fruitful in at least three contexts: (1) interactions of professionals and patients; (2) culture; and (3) public policy in a liberal, pluralistic democracy. Attending to religious perspectives and listening to religious voices can enrich bioethics in these three contexts, which are, of course, not totally separate from each other. My call for increased attention to religious and theological perspectives is a call for dialogue, not a call for granting such perspectives primacy in bioethics. In various ways, they may provide insights and even wisdom. Religiously oriented and theological bioethicists also stand to gain from this dialogue, not least because they must respond to the demand for greater clarity and precision and consider possible points of convergence with secular views. And I believe they can do so without diminishing the power of religious and theological perspectives to illuminate bioethics in significant ways

# Notes and References

[1] I will proceed with only rough characterizations of these categories. Religion includes a wide range of beliefs, norms, symbols, rituals, etc., which are involved in a community's (and individual's) response to the sacred, holy, transcendent (on a substantive definition of religion) or which express ultimate concern (on a formal definition). The disciplines that make up the broad area of religious studies examine the phenomena of religion through various methodologies. Whereas religion is closer to practice, theology is a reflective activity, often but not necessarily systematic in character.

[2] *See* the summary of recent discussions by Robert M. Veatch. (1995) Medical codes and oaths: history, in *Encylopedia of Bioethics*. (Warren T. Reich, ed. in chief), MacMillan Library Reference, New York, NY, 3: p. 1420.

[3] Albert R. Jonsen. (1994) *The Birth of Bioethics*. Oxford University Press, New York, NY, p. 34.

[4] LeRoy Walters. (1985) Religion and the renaissance of medical ethics in the United States: 1965–1975, in *Theology and Bioethics: Exploring Foundations and Frontiers*. (Earl E. Shelp, ed.), D. Reidel Publishing Company, Dordrecht, Holland, pp. 3–16.

[5] Daniel Callahan. (1990) Religion and the Secularization of Bioethics.*Hastings Center Report*, 20 (Jul/Aug), p. 2.

[6] Allen Verhey and Stephen E. Lammers. (1993) Introduction: rediscovering religious traditions in medical ethics, in *Theological Voices in Medical Ethics*. (Allen Verhey and Stephen E. Lammers, eds.), William B. Eerdmans Publishing Company, Grand Rapids, MI, p. 3.

[7]Carla M. Messikomer, Renee C. Fox, and Judith P. Swazey. (2001) The Presence and Influence of Religion in American Bioethics. *Perspectives in Biology and Medicine.* 44: No. 4 (Autumn), pp. 485–508. All subsequent references to their work are to this article.

[8]Another important recent work argues that bioethics, particularly concerning genetic engineering, moved from thick religious/theological concerns (under substantive rationality) to thin principles. It too fails to offer a clear definition of religion and theology, and thus to identify clear criteria for classifying thinkers. *See* John H. Evans. (2002) *Playing God? Human Genetic Engineering and the Rationalization of Public Bioethical Debates.* University of Chicago Press, Chicago, IL.

[9]*See* Tom L. Beauchamp and James F. Childress. (2001) *Principles of Biomedical Ethics.* (5th edition),Oxford University Press, New York, NY. For several articles that consider whether particular religious traditions have, and flesh out, these intermediate principles, *see* Ranaan Gillon, ed. (1994) *Principles of Health Care Ethics.* John Wiley and Sons, Chichester, England.

[10]*See* Peggy DesAutels, Margaret P. Battin, and Larry May. (1999) *Praying for a Cure: When Medical and Religious Practices Conflict* . Rowman and Littlefield, Lanham, MD.

[11]For confusions of Jehovah's Witnesses with Christian Scientists, *see* Gerald Dworkin. (1971) Paternalism, in *Morality and the Law.* (Richard Wasserstrom, ed.), Wadsworth Publishing Co., Belmont, CA; and Michael Bayles. (1978) Paternalism, in *Principles of Legislation.* Wayne State University Press, Detroit, MI .

[12]Dena Davis. (1994) It Ain't Necessarily So: Clinicians, Bioethics, and Religious Studies. *J. Clin. Ethics* 5(4): 315–319.

[13]Joseph A. Carese and Lorna A. Rhodes. (1995) Western Bioethics on the Navajo Reservation: Benefit or Harm? *J. Am. Med. Assoc.* 274 (Sept 13): 826–829.

[14]A.L. Kroeber and Clyde Kluckhon. (1952) *Culture: A Critical Review of Concepts and Definitions.* Vintage, New York, NY, p. 66.

[15]*See* James Davison Hunter. (1991) *Culture Wars: The Struggle to Define America.* Basic Books, New York, NY.

[16]James M. Gustafson. (1978) Theology Confronts Technology and the Life Sciences. *Commonweal,* 105 (June 16): p. 392.

[17]Paul Ramsey. (1970) *The Patient as Person.* Yale University Press, New Haven, CT.

[18]*Ibid.*, pp. xii–xiii.

[19]Paul Ramsey. (1978) *Ethics at the Edges of Life: Medical and Legal Intersections.* Yale University Press, New Haven, CT, p. xiv.

[20]Stanley Hauerwas. (1996) How christian ethics became medical ethics: the case of Paul Ramsey, in *Religion and Medical Ethics: Looking Back, Looking Forward.* (Allen Verhey, ed.), William B. Eerdmans Publishing Company, Grand Rapids, MI, p. 79. For another very valuable interpretation of Ramsey's medical ethics, *see* David H. Smith. (1993) On Paul Ramsey: a covenant-centered ethic for medicine, in *Theological Voices in Medical Ethics.* (Allen Verhey, ed.), pp. 7–29.

[21]William F. May. (1985) Religious Justifications for Donating Body Parts. *Hastings Center Report*, 15 (Feb), pp. 38–42.

[22]Joel Feinberg. (1985) The Mistreatment of Dead Bodies. *Hastings Center Report*, 15 (Feb), pp. 31–37.

[23]Gilbert Meilaender. (1993) On William F. May: corrected vision for medical ethics, in *Theological Voices in Medical Ethics*, pp. 110,125.

[24]Gilbert Meilaender. (1997) Remarks on Human Cloning to the National Bioethics Advisory Commission. *BioLaw* 2(6): S114–118.

[25]Richard McCormick, S.J. (1989) Theology and Bioethics. *Hastings Center Report*, 19:3, pp. 5–10.

[26]*See* H. Tristram Engelhardt, Jr. (2000) *The Foundations of Christian Bioethics.* Swets & Zeitlinger Publishers, Lisse, The Netherlands; and Stanley Hauerwas. (2001) *The Hauerwas Reader.* (John Berkman and Michael Cartwright, eds.), Duke University Press, Durhman, NC, especially Chapters 27–31.

[27]R.C. Lewontin. (1997) The Confusion over Cloning. *New York Review of Books*, Oct 23.

[28]James M. Gustafson. (1990) Moral Discourse about Medicine: A Variety of Forms. *J. Med. Philos.* 15(2): pp. 127–141.

[29]*See* Joel James Shuman. (1999) *Body of Compassion: Ethics, Medicine, and the Church.* The Westview Press, Boulder, CO. *See also* Courtney S. Campbell. (1999) Bearing witness: religious resistance and meaning, in *Notes from a Narrow Ridge: Religion and Bioethics.* (Dena S. Davis and Laurie Zoloth, eds.), University Publishing Group, Hagerstown, MD, pp. 21–48.

[30]For a discussion of a New Jersey statute that exempts certain religiously based conscientious objections to the brain death stan-

dard, *see* Robert Olick. (1991) Brain Death, Religious Freedom, and Public Policy. *Kennedy Institute of Ethics Journal* 1(Jun): pp. 275–288.

[31]*See* James F. Childress. (1999) Religion, morality, and public policy: the controversy about human cloning, in *Notes from a Narrow Ridge: Religion and Bioethics.* (Davis and Zoloth, eds.), pp. 65–85, from which I have drawn some ideas, formulations, and a few paragraphs.

[32]National Bioethics Advisory Commission. (1997) *Cloning Human Beings: Report and Recommendations of the National Bioethics Advisory Commission.* National Bioethics Advisory Commission, Rockville, MD, Vol. I: pp. 7,8,39,40.

[33]John Rawls. (1993) *Political Liberalism.* Columbia University Press, New York, NY, 220, Note 42.

[34]Michael J. Perry. (1997) *Religion in Politics: Constitutional and Moral Perspectives.* Oxford University Press, New York, NY, p. 45.

[35]Michael Sandel. (1995) Political Liberalism: Religion and Public Reason (a symposium). *Religion and Values in Public Life* 3(4): p. 3.

[36]For a somewhat similar view, *see* Robert Audi. (1997) Liberal democracy and the place of religion in politics, in *Religion in the Public Square: The Place of Religious Convictions in Political Debate.* (Robert Audi and Nicholas Wolsterdorff, eds.), Rowman and Littlefield, Lanham, MD.

[37]Carla M. Messikomer, Renee C. Fox, and Judith P. Swazey, *supra note 7*, p. 505.

[38]*See* John H. Evans, *supra note 8*.

# 3

# Medicine and Bioethics

## *Howard Brody*

Medical practitioners interested in ethics have tended to view bioethics and medical ethics as virtually synonymous terms. It might therefore seem impossible to discuss the impact of bioethics on medicine, and of medicine on bioethics, without discussing the whole of bioethics within this one chapter. This, I hope, excuses some selection to narrow the scope of inquiry.

I first address the impact bioethics has had on medical practice and second, some of the important medical–ethical issues that may as yet have been inadequately addressed by bioethics.

The debate I seek most to avoid is that of whether bioethics consultation in health care settings is better done by physicians or by other professionals. Some strong claims have been made over the years regarding the need for specific clinical skills if one is to perform bioethics consultation adequately. Sometimes these claims have been disguised attempts to state simply that only physicians are really qualified to do ethics consultation. Indeed, in some instances, it is barely disguised, such as when it was proposed that the bioethics consultation should include a physical examination of the patient. I prefer to regard this entire debate, which seems in retrospect quite sterile, as fundamentally a turf

From: *The Nature and Prospect of Bioethics: Interdisciplinary Perspectives*
Edited by: F. G. Miller, J. C. Fletcher, and J. M. Humber
© Humana Press Inc., Totowa, NJ

battle, which tended to ignore, if not dismiss the interdisciplinary nature of bioethics. As this volume takes the interdisciplinarity of bioethics for granted, I hope no more need be said on this point. Suffice it to say that nothing in this chapter should be taken to imply that physicians are better able to "do bioethics," whatever that might consist of, than other professionals with the proper training.

## Impact of Bioethics on Medicine

Assuming that what we now call bioethics came into being during the past thirty to forty years, a historical analysis of medicine during that period is necessary to begin to assess the impact of bioethics on medicine. Fortunately, such a study is at hand in David Rothman's *Strangers at the Bedside (1)*. I find the book quite persuasive in its key points, but also in need of one correction.

Rothman cleverly chose his title to do double duty in support of his thesis. The rise of bioethics during the key decade of 1966 to 1976—corresponding to the period between Henry Beecher's exposé of unethical medical research in the United States *(2)*, and the Karen Quinlan case—resulted in non-physician "strangers" appearing at the patient's bedside to help physicians with perplexing ethical problems, especially those created by new technologies, such as ventilators and organ transplants. But Rothman finds it inconceivable that these strangers would have found a place at the patient's bedside had not medicine already changed in a fundamental way in the preceding decades. Rothman marshals evidence to show that it is much more likely that the physician responsible for caring for the patient when the ethical dilemma was identified would herself be a stranger to the patient, in contrast with earlier times when the responsible physician would more likely have been a trusted retainer who had served the patient for some time, and who might have lived just down the street. If the physician at the bedside was himself a stranger, there because of special expertise he possessed and not because of any prior contact with the patient, he could hardly complain if he were to be joined by other strangers, with different but relevant special expertise.

This historical observation sets up some firm conclusions as to the form that bioethics in medicine would then take. One would naturally expect what Robert Burt called an ethic of "strangers," focusing on rights and principles *(3)*. Rights is the natural language to use when the interests of strangers collide, and principles is the natural way to characterize the special, impersonal expertise that the bioethics strangers might bring to the clinical setting. The further result is that today, patient autonomy is part of the language and the mindset of medicine. There is, to be sure, a spectrum—some physicians grudgingly recognize autonomy as a foreign importation invented by malpractice attorneys, whereas others actively embrace autonomy as not only a new and better way to practice medicine, but also as a way to make patients more effective in promoting their own health. It is almost unimaginable to find physicians in the United States today who are not familiar with the concept of patient autonomy, which was surely not the case twenty years ago.

When writing a history of a period in which the principal actors are all still alive, one can be sure to receive pointed criticism from many who remember events differently. I think some of the criticism Rothman's analysis has received needs to be taken with a grain of salt, and I am conscious of the danger of offering a rebuttal that I can defend only with general recollections and no hard data. Nonetheless, my own impression, based on having entered the field of bioethics as a medical student in 1972, is that Rothman omits an important factor. He implies that bioethics invaded medical space, and that most doctors wished to repel the invader but could not succeed for a variety of political and social reasons. He omits, both in my view and in that of other reviewers of his book, the extent to which a number of prominent physicians actively invited fledgling bioethicists into the clinical world.

These "turncoat" physicians were perhaps few in numbers, but many were quite prominent and influential. (To name one example, Tom Hunter, long-time dean of the University of Virginia Medical School, invited Joseph Fletcher, the theologically trained ethicist, to become a visiting professor.) If one looks at the general status of ethical studies of the professions in the United States in this era, medicine (and other health professions like nursing and dentistry) is head-and-shoulders above law, business, and journalism in its eagerness to subject its practices to searching ethical scrutiny and to invite outsiders to teach future practitioners about ethics.

Why should American medicine, or at least an influential segment within it, turn to the bioethics "strangers" in this way? The standard answer is that new technology forced the issue, by raising the question asked at the beginning of almost every book and article on bioethics in the 1960s and early 1970s: "Just because we can do it, should we do it?" But this view, although containing some truth, is far too simple. Such bioethics warhorses as truth-telling, informed consent, and euthanasia have histories going back decades and even centuries before the invention of modern medical gadgetry. Rothman himself locates the irritant grain of sand that produced the pearl of bioethics in the practices of research on human subjects on both sides during World War II, a low-tech era by today's standards.

Gerald Osborn developed a more plausible thesis based on a legacy of oral histories elicited from many early bioethics leaders *(4)*. Osborn found that most of them, physicians included, occupied the left-of-center portion of the political spectrum of the 1960s. They would thus be in general sympathy with those who challenged traditional expertise and authority in the anti-war, environmental, and civil rights movements. From within medicine, they would seek to position themselves on the side of those who would question the tradition of "doctor knows best." Thus, these physicians would see theologians, philosophers, and lawyers doing pioneer work in bioethics as potential allies rather than as threats.

A further answer is explored by some other oral histories obtained by anthropologist Sharon Kaufman from eight prominent physicians who were nearing or reaching the twilight of their medical careers during the period 1966 to 1976 *(5)*. Kaufman shows these physicians as having witnessed profound changes in the practice of medicine during their professional lifetimes. These physicians had all allied themselves early on with what had then appeared to be the forces shaping the future of medicine; and yet they still failed to anticipate these changes. Most had subspecialized and had undertaken medical research. They did so with the unspoken assumption that medicine possessed a solid core of clinical wisdom and commitment to the well-being of the patient, and that the new science would supplement and expand on that desirable core. But, by the end of their careers, they found the world of research and subspecialization reshaping medicine in ways they saw as threatening some fundamental aspects of medicine's moral foundation.

I would speculate a bit beyond Kaufman's data to suggest what might have disturbed these senior physicians of the 1960s and 1970s. First, they may have found the new science of medicine creating and compounding uncertainty. As students, residents, and junior academicians, they had probably assumed that the new science would gradually, by accretion, replace ignorance with knowledge. However, by the 1970s the adage for new medical students had become, "Whatever they teach you in school, fifty percent will be wrong in ten years; the problem is you won't know which fifty percent." Instead of a tree with solid roots growing new branches and foliage, medicine sometimes seemed more like tumbleweed, blown by the wind first this way, then that way. Many avoided confronting this disquieting uncertainty by subspecializing, hoping that so long as one marked out a sufficiently narrow band of practice as one's domain, a high level of certainty could still prevail. These senior physicians, however, did not have this mindset to fall back on. They had trained in an era when one first became a solid generalist, and only then did one subspecialize. Despite their own specific areas of research and clinical expertise, they were prone to view the world as generalists; and as generalists much more inclined than subspecialists to discern, and even to embrace, medicine's inherent uncertainty.

Second, and closely related, these senior physicians saw the field being taken over by a new species of physician. The seniors, and their teachers before them, had adhered to the traditional model of first becoming a well-rounded generalist and only then subspecializing. But increasingly, medical training seemed to produce fast-track subspecialists who never bothered with general medicine before learning their specific line of work. Moreover, as generalism was now widely sneered at, there seemed little hope in the future that any new physicians would be trained as these senior physicians had been. This meant, besides the tunnel vision that produced a spurious sense of medical certainty, that the new subspecialists would seldom consider the personal relationship between the physician and the patient as something important for medical practice, worth preserving if it was under fire.

If this account is plausible, it follows that a number of thoughtful, senior physicians in the 1960s and 1970s might have felt a deep level of moral concern about the future of medicine— and a good deal of distrust in the ability of their younger colleagues to set matters right. They might then have been quite sympathetic to the overtures of the new field of bioethics, and might have welcomed it as a needed corrective within medicine— even if they had little understanding of what exactly bioethics consisted of, or where it might lead.

## An Underaddressed Issue: Physician Integrity

If bioethics has had notable successes in influencing medical practice, are there any areas that remain relatively unexplored? I suggest two; and both relate to the dominance, until recently, of "principlist" approaches in US medical ethics. I lack the space here for a full discussion of the pros and cons of principles as opposed to casuistry, virtue, narrative, and other approaches that have recently attracted attention *(6)*. All I aim to conclude, in order to launch further discussion, is that bioethics has tended to focus on rules the good physician should follow, to the exclusion of what sort of person the good physician should be (or should try to become) *(7)*.

Also, the rules that have received the most attention have been patient-centered (most obviously and notably the rule of respect for patient autonomy). This gives rise to two apparently contradictory reactions. The first is that this focus is appropriate, as medicine is, by its nature, a profession of service to the patient. To start medical ethics with a physician-centered approach would therefore deny at the outset its moral reason for being. But the second reaction is that physicians, after all, come to their work in the role of physician and not in the role of patient. To focus solely on the patient, and to treat the physician *merely* as someone who will serve the patient well if only he or she adheres to the right principles, may thus run the risk of obscuring important aspects of medical ethics.

The first issue that may have been obscured in this way is professional integrity, which is closely related to the internal morality of medicine. Frank Miller and I have tried to explicate this concept, and have elsewhere applied it to three specific bio-ethical issues: assisted death *(8)*; futile therapy *(9)*; and physician obligations within managed care *(10)*. What is perhaps most notable for this discussion is the fact that Miller and I were forced to begin almost *de novo*, and to develop our own account of pro-fessional integrity. Although traditional codes of medical ethics might be viewed as focusing on physician integrity, sometimes to the exclusion of patients' rights, the bioethics literature of the past thirty years gave us very little raw material from which to fashion an account of professional integrity.

We concluded by arguing that the physician of integrity pur-sues one or more of a number of goals in medical practice and he

Table 1
Elements of Professional Integrity

---

Legitimate Goals of Medical Practice

---

1. Reassuring the "worried well" who have no disease or injury.
2. Diagnosing the disease or injury.
3. Helping the patient to understand the disease, its prognosis, and its effects on his or her life.
4. Preventing disease or injury if possible.
5. Curing the disease or repairing the injury if possible.
6. Lessening the pain or disability caused by the disease or injury.
7. Helping the patient to live with whatever pain or disability cannot be prevented.
8. When all else fails, helping the patient to die with dignity and peace.

---

Ethically Acceptable Means of Medical Practice

---

1. The physician must employ technical competence in practice.
2. The physician must honestly portray medical knowledge and skill with both the patient and the general public, and avoid any sort of fraud or misrepresentation.
3. The physician must avoid harming the patient in any way that is out of proportion to expected benefit, and must seek to minimize the indignity and the invasion of privacy involved in medical examinations and procedures.
4. The physician must maintain fidelity to the interests of the individual patient.

---

Adapted from Miller and Brody *(10)*.

does so within the bounds set by ethically acceptable means of practice, as summarized in Table 1.

An important implication of this account is that medicine is a multifaceted activity by its nature. No statement of a single goal can adequately characterize medical practice. For example, those who claim that euthanasia can never be ethical, because the physician must always be a healer, may be guilty of misrepresenting medical practice as if it can be defined by a single goal.

Another important implication is that professional integrity might constitute an important (even if limited) check on the exercise of patient autonomy. In most cases that arise within medical practice, respect for patient autonomy will dictate the best course of action. In a subset of cases, however, there might be a direct conflict between what the informed patient autonomously requests and what the physician of integrity feels able to perform. A patient who is physically able to return to work, but requests an extended medical leave, is one simple example. In these cases, professional integrity may dictate that the patient be refused what she requests. This is not to deny the general moral force of patient autonomy. Rather, it reflects the assumption that the patient comes to the physician for a very specific reason—to receive medical care. And the complex, evolving nature of the practice of medicine implies that neither the individual patient, nor the individual physician, may unilaterally redefine what "good" practice consists of to suit each one's own purposes. At some point, the ethical physician must fall back on some agreed on (even if poorly articulated) standards, and say, "If you ask me to do what you want, you may be asking for something that will advance your personal interests in some way, but it is not medical care, or at least not acceptable medical care."

One point here needs special emphasis: Physicians who feel beleaguered by the emphasis on patient autonomy have often asked, "What about physicians? Aren't we people too? Doesn't *my* autonomy count for something?" This in turn requires a careful distinction between *personal* and *professional integrity*. The physician, as a person of integrity deserving equal respect as a moral agent, should not be required to do something that violates his or her personal moral code. But that assertion rests on individual autonomy in a way that an appeal to professional integrity does not.

The example that may perhaps most easily illustrate the difference is abortion. It is generally accepted that a physician whose personal religious or philosophical convictions prohibit abortion need not participate if patients request it. The reasoning here is covered by an appeal to individual autonomy; to require this physician to perform an abortion is to deny his or her moral agency and to treat the physician as a mere means to the patient's satisfaction. But such an appeal is different from an appeal to professional integrity. This latter appeal would claim not only that this physician cannot perform the abortion; it would also claim that *no ethical physician* could perform an abortion. And I assert, on both empirical and ethical grounds, that this latter claim could not be sustained in US medical practice today.

Thus, appeals to professional integrity actually constrain the physician's individual autonomy. To become a physician of integrity, one must learn the ethical standards and the internal moral system that define medicine. That requires subsuming one's own choices and inclinations, to a large degree, to bring one's practice into conformity with that of other physicians. To a large extent, patients trust physicians (many of whom they may never have met prior to the present encounter) precisely because of the predictability of "physicianly" behavior. The patient, for example, does not generally feel the need to ask, "I know that other physicians feel bound to maintain patient confidentiality; do you also feel this way?" The patient simply assumes that the title "physician" entails commitment to some set of basic norms of practice. And this can be true only if physicians are willing, especially in the course of training, to forgo considerable exercise of individual autonomy.

This account of professional integrity posits a level of ethical analysis in between one's personal integrity, which may be idiosyncratic, and the general ethical rules and principles applicable to the entire community. That is, medicine is presumed to have its own internal moral code growing out of the nature of the practice, which is binding on physicians but not on non-physicians. There are several objections that could be offered to this formulation; but one which helps further to clarify the assertion is that this unduly privileges traditional medical practices and behavior. One reason why bioethics did not focus especially on physician integrity in the 1960s and 1970s, and instead paid much more attention to patient autonomy, was that traditional physician practices were indefensibly paternalistic. Had bioethics focused on professional integrity (this objection goes), the result would have been inevitably to reinforce, rather than to call into question, these paternalistic habits.

For this very reason, the internal morality of medicine must be viewed as a responsive tradition, always in a creative tension between its historical roots and the changing social environment within which medicine is practiced *(10,11)*. To defend paternalism in US medical practice in the 1960s, for example, would have been to focus solely on the "tradition" part without paying any attention to the "responsive" part. By contrast, the Nazi physicians stand condemned of totally forgetting their tradition (as well as ethics more generally) in the name of responding to the dominant political climate of Germany in the 1930s. There is obviously no simple formula for how much "tradition" and how much "responsiveness" would constitute the correct mix. Some Hippocratic traditionalists will no doubt reject this picture because they prefer to regard medical ethics as something that can simply be read off the historical record. I see no realistic choice but to accept the inherent and inevitable tension of the notion of a "responsive tradition," even if it leads to some degree of ethical controversy and uncertainty.

Recent changes in how US health care is financed drive home the need for some conception of professional integrity and of a moral core that defines medical practice. What are we to say, for example, to a physician who states, "Under the old fee-for-service system, I made more money if I did more for the patient. But our society has decreed that such a system is to be replaced with a managed-care-dominated system, in which the financial incentives favor doing less. Because I am no longer able to practice in the way I used to, I might as well give up any pretense of serving the patient, and figure out how to work today's system to make as much money as possible before I retire"? Leaving aside for the moment the question of what we might think about the ethics of managed care, how would we characterize this physician's resolution?

If some critics are correct, and there is no such thing as an internal morality of medicine, then we are left only with two questions. First, does this behavior violate this particular physician's own personal sense of moral integrity; and second, is such behavior wrong according to the rules and principles that ought to govern all people in society? We have assumed that there is no violation here of the physician's personal values; and if this physician proceeds warily enough, assuring, for instance, informed consent for all the "consumers" with whom he interacts, there need be no violation of general ethical norms either. (After all, making a lot of money is supposed to be a good thing in a capitalist society.) I submit, however, that these two questions omit a good deal that is ethically important about medical practice. I think the minimum statement one would wish to make is that this physician has fallen short of one of the important moral ideals of the practice of medicine. One could make the statement even stronger by saying that this physician is woefully lacking in professional integrity and perhaps even go so far as to say that, in an important sense, he has ceased to practice *medicine* at all.

## An Underaddressed Issue: Physician Self-Interest

The second area that has escaped serious attention in modern bioethics follows closely on the first. If the virtue and character of the physician is dismissed in favor of a focus on rules and principles, then it seems inevitable that this second issue will receive scant attention as well.

This issue was squarely identified by Albert Jonsen, who wrote that a central (if not *the* central) ethical problem in medicine is the need to balance the physician's altruistic duty to serve the patient with the physician's legitimate self-interest *(12)*. Once stated, this issue seems so self-evident that one wonders why Jonsen felt any need to explicate it. But one then may look in vain for careful discussions or elaborations of this ethical conflict in the vast majority of textbooks and articles on bioethics.

The "majority view" on this subject, which is the same as the duty of fidelity listed in Table 1, is simply that the physician is morally obligated above all else to serve the needs of the patient. Most works on bioethics either state this as an absolute, unqualified obligation; or else hedge considerably on what circumstances might override this duty in actual practice. Some reflection, however, reveals that this is not how most physicians actually behave.

Consider an example that may at first seem contrived and unrealistic, but that will, I hope, prove instructive on further dissection. A physician decides that a patient requires a liver transplant to survive. The transplant will cost perhaps $150,000, and is not covered by the patient's insurance. The patient and the patient's family have tried hard to raise money in every way they can, but are still $40,000 short of their goal. It so happens that this physician has saved $40,000 in a college fund for his or her children (who are not yet of college age). Should the physician use these savings to pay for the transplant?

The first thing to notice here is that many influential physicians have argued that their central moral duty is to do what is best for the individual patient, regardless of cost *(13)*. This statement is repeated particularly in the face of pressures for cost containment in managed care today. The phrase, *"regardless of cost,"* rolls off one's tongue quite easily when it is assumed that the costs are to be borne by *someone else* (presumably the managed care plan, or the impersonal "system.") The phrase sticks in the physician's throat, however, when it is contemplated that the physician might personally have to assume some of that cost. But if this is an *absolute* moral duty, why should it not follow logically that this physician should donate the money toward the transplant?

The next likely objection is that stated this way, the duty is obviously unrealistic. After this patient either gets his transplant or dies, an indefinite number of patients will come forth, each needing some expensive care; there is no practical way in which this one physician could pay for all their unmet needs. But this rejoinder runs afoul of precisely the error which the "absolute fidelity" people accuse managed care systems of ignoring—focusing on what is best for a group or population of patients instead of on the needs of the one patient who is now before us. In this view, the physician is false to his or her ethical obligations as soon as he or she starts to think of trade-offs within populations; to maintain an ethical compass, the physician must think only of the duty to the individual patient, one at a time. Very well, this individual patient is here now; the patient needs $40,000, and the physician has $40,000. If the physician helps to pay for the transplant and another patient comes tomorrow who needs $40,000 in uninsured treatment, then the physician won't have the $40,000 and will be unable to help. According to the "absolute fidelity" point of view, there should be nothing wrong with that; and the fact that the next patient might show up tomorrow does nothing to mitigate the duty to spend the $40,000 today ("regardless of cost.")

A final objection would be that I have totally misunderstood the "absolute fidelity" statement as if it were a rule, whereas it is actually a statement of a moral ideal. A moral ideal could provide valuable guidance even if in practice no physician ever fully attains it. This makes perfectly good sense and would be persuasive, if only the dominant focus of bioethics in the past thirty years had been on virtue and character. However, because bioethics has been overwhelmingly about rules and principles, which are supposed to provide guidance in practical dilemmas, it seems odd to claim suddenly that this often repeated formula was not supposed to be a rule after all. Physicians who have used this formula to condemn gatekeeping in managed care certainly view it as a moral rule and not as a moral ideal.

If physicians in practice do not follow the "absolute fidelity" rule, what moral principle do they adhere to? First, note that some physicians do in fact follow the rule—they go into mission-ary or inner-city practices and cheerfully impoverish themselves so that they can serve the neediest in society. But most other physicians regard their behavior as supererogatory rather than obligatory. The other physicians adhere to a rule of placing the patient's welfare on a higher priority plane than would be usual in the world of business, for instance, and indeed accepting notable sacrifices in the name of the patient's good—especially when it comes to training for many years at low pay and working long hours. But this rule always entails that there comes a point at which the physician's self-interest will take priority. The physician will, in general, not give up a certain comfortable level of income; or a desirable place of residence; or excellent schooling for his or her children, even if doing so could better serve the welfare of some patients.

If this is the actual, real-life moral rule by which physicians live, then why has virtually no attention been paid to it in the bioethics literature? The answer, I think, differs for physician and non-physician bioethicists. The physicians have, probably without realizing it, fallen into the same trap as organizations like the American Medical Association (AMA). Those trying to preserve medicine's prestige, especially in an era in which authority groups of all kinds are under fire, find the "absolute fidelity, regardless of cost" rhetoric very useful. A candid statement of a need to balance altruism against self-interest would hardly be suitable as a public-relations strategy.

In this matter, it should be noted that the AMA actually deviates from its own historical–ethical roots. The original code of ethics for the AMA, following the example of Thomas Percival's earlier English code, listed obligations owed to physicians by the individual patient and by society at large, alongside the obligations owed by the physician to the patient and to the community. Today we tend to regard those portions of the old code of ethics as quaint anachronisms, rather than as important gems of ethical wisdom.

Non-physician bioethicists, on the other hand, have been under pressure to criticize medical practices and norms while not appearing to attack physicians in any sort of *personal* way. Our society views how much money one makes and what one does with it as an intensely personal matter. Thus, addressing the physician's income as a bioethical issue would appear simply as jealous "doctor-bashing" rather than as serious ethical scholarship. And so few in bioethics have dared even to mention physician income as an ethical issue, much less propose principles by which one could judge what level of remuneration is *owed* physicians as a matter of fairness *(14,15)*.

An unfortunate consequence of ignoring the tension between altruism and legitimate self-interest, is that despite all the positive influence bioethics has had on medical practice, physicians at some level are tempted not to take bioethics seriously. As long as bioethics seems to expound a principle that physicians know that they cannot follow in their everyday lives, bioethics becomes the Sunday sermon that sounds nice only as long as one is free to act otherwise all the rest of the week. This, I submit, is too high a price to pay for being polite to physicians by not putting their personal finances under the bioethical lens.

## Conclusion

Bioethics arrived some thirty years ago and found physicians too arrogant and unreflective in how they used their power over patients. Bioethics set about to correct this, and found important allies among physicians who were themselves concerned about the direction in which their profession was moving. Today, bioethics has made great strides and has had a profound influence on medical practice, at least in the United States. To move forward in medicine, ironically, bioethics must now turn at least some of its attention away from the patient and divert it back to the physician.

# References

[1]D. J. Rothman. (1991) *Strangers at the bedside: a history of how law and bioethics transformed medical decision making.* Basic Books, New York, NY.

[2]H. K. Beecher. (1966) Ethics and clinical research. *N. Engl. J. Med.* 274: 1354–1360.

[3]R. A. Burt. (1979) *Taking Care of Strangers: The Rule of Law in Doctor–Patient Relations.* Free Press, New York.

[4]G. G. Osborn. (1985) *The Doctor's Dilemma: Ethics in Twentieth Century Medicine [Dissertation].* University of Cambridge.

[5]S. R. Kaufman. (1993) *The Healer's Tale: Transforming Medicine and Culture.* University of Wisconsin Press, Madison, WI.

[6]H. Brody. (1992) *The Healer's Power.* Yale University Press, New Haven.

[7]D. G. Smith and L. Newton. (1984) Physician and Patient: Respect for Mutuality. *Theoret. Med.* 5: 43–60.

[8]F. G. Miller and H. Brody. (1995) Professional integrity and physician-assisted death. *Hastings Center Report* 25(3): 8–17.

[9]H. Brody. (1997) Medical futility: a useful concept, in *Medical Futility and the Evaluation of Life-Sustaining Interventions.* (M. B. Zucker and H.D. Zucker, eds.) Cambridge University Press, Cambridge, pp. 1–14.

[10]H. Brody and F. G. Miller. (1998) The Internal Morality of Medicine: Explication and Application to Managed Care. *J. Med. Philos.*

[11]R. Vance. (1985) Medicine as Dependent Tradition: Historical and Ethical Reflections. *Perspect. Biol. Med.* 28: 282–302.

[12]A. R. Jonsen. (1990) *The New Medicine and the Old Ethics.* Harvard University Press, Cambridge, MA.

[13]N. G. Levinsky. (1984) The Doctor's Master. *N. Engl. J. Med.* 311: 1573–1575.

[14]P. T. Menzel. (1983) *Medical Costs, Moral Choices.* Yale University Press, New Haven, CT, pp. 213–229.

[15]U. E. Reinhardt. (1987) Resource Allocation in Health Care: The Allocation of Lifestyles to Providers. *Milbank Q.* 65: 153–176.

# 4

## When Policy Analysis Is Carried Out in Public

### Some Lessons for Bioethics from NBAC's Experience

*Eric M. Meslin*

Policy discussions involving bioethical topics most often capture the attention of the public, health professionals, and elected officials when the topics are controversial and do not readily give rise to clear-cut solutions. Much has been written about the role of ethics, ethical principles, and ethical theory in the formation of public policy, and especially policy involving bioethical subjects such as end-of-life care, allocation of high technology resources in hospitals, reproductive decision making, cloning, and stem cell research. Those familiar with the bioethics literature will recognize two broad types of discussions. The first are discussions that tend to illuminate and resolve particular conceptual problems, such as the appropriate definition of terms: moral rights, capacity to consent, or just allocation of resources. The second type of discussion focuses on specific prescriptive tasks: regulations or guidelines for the protection of human subjects in

From: *The Nature and Prospect of Bioethics: Interdisciplinary Perspectives*
Edited by: F. G. Miller, J. C. Fletcher, and J. M. Humber
© Humana Press Inc., Totowa, NJ

federally sponsored research; legislation to enhance access to health care; or plans to allocate organs for transplantation.

In the introduction to their edited volume *Encyclopedia of US Biomedical Policy*, Blank and Merrick *(1)* argue that although bioethics and policy analysis overlap in certain ways, they have different emphases. Biomedical policy has three dimensions:

> First, decisions must be made concerning the research and development of technologies. Because a substantial proportion of medical research is funded either directly or indirectly with public funds, it is important that public input be included at this stage. ...The second policy dimension relates to the individual use of technologies once they are available. Although direct government intrusion into individual decision making in health care has, until recently, been limited, the government does have at its disposal an array of more or less implicit devices to encourage or discourage individual use. ...The third dimension of biomedical policy centers on the aggregate consequences of widespread application of a technology. ...Policy planners must account for these potential pressures on the basic structures and patterns of society and decide whether provision of such [technology] choices is desirable. (p. xiii)

It is no surprise then, that bioethics policy discussion has increasingly become a part of the fabric of public discourse. One need look no further than August 9, 2001, the day US President George W. Bush used his first national television appearance as president to announce his policy about the use of federal funds for research involving human embryonic stem cells:

> Embryonic stem cell research offers both great promise and great peril. So I have decided we must proceed with great care. As a result of private research, more than sixty genetically diverse stem cell lines already exist. They were created from embryos that have already been destroyed, and

they have the ability to regenerate themselves indefinitely, creating ongoing opportunities for research. I have concluded that we should allow federal funds to be used for research on these existing stem cell lines, where the life and death decision has already been made.

Leaving aside for the moment whether this policy is a good one, whether it is based on sound ethical reasoning, or what the consequences are for science and society, consider what it means for a recently elected president, in a time of peace, to use the airtime of all the major television networks to address the nation—not about the federal budget, tax relief, education, the national defense, or social security—but about a scientific discovery, and the ethical implications of that discovery for the country. Many people, including some who, days before had never heard of a "stem cell" began to formulate opinions or revisit existing beliefs about the moral status of the human embryo and the role of federal government and the private sector in overseeing this research.

## The Process of Policy Construction as the Case

Bioethical inquiry can be most helpful when it focuses its attention on a case. Cases provide analysts with the context, history, facts, and content for ethical analysis that follows. The strength of this form of analysis, casuistry, is in the inductive power the analysis might have. For this reason, case study is particularly well suited to individual problems in health care, where patients, families, and practitioners provide the context and are the beneficiaries of the analysis *(2)*. However, policy problems are not merely larger versions of clinical issues. Public policies focus on the interests and well being of groups of individuals, and society as a whole, and case study is used in this. The Institute of Medicine (IOM) report, *Biomedical Politics*, used historical case study for a specific purpose:

Case study excuses us from addressing issues on the level
of high theory and general principles by injecting a large
measure of pragmatism. Indeed, democracy in America
seems to preclude the development of fixed universal laws
and immutable principles, and attempts to set policy on such
assumptions tend to create irresoluble debate. (ref. *3*, p. 4)

This pragmatism is in contrast with what Joel Feinberg characterized as the ideal method in which public policy is informed by
ethical analysis *(4)*:

It is convenient to think of these problems as questions for
some hypothetical and abstract political body. An answer to
the question of when liberty should be limited or how wealth
ideally should be distributed, for example, could be used to
guide not only moralists, but also legislators and judges
towards reasonable decisions in particular cases where
interests, rules, or the liberties of different parties appears
to conflict. ...We must think of the ideal legislator as somewhat abstracted from the full legislative context, in that he
is free to appeal directly to the public interest unencumbered
by the need to please voters, to make "deals" with colleagues, or any other merely "political" considerations. (ref.
*4*, pp. 2–3)

Seen in this contrast, we are left with an important choice:
should policy-makers aspire to pragmatism or idealism?
Feinberg's view idealizes the goal of public policy at the expense
of the reality of the process. Public policy occurs in real time,
with fallible personalities, and all the encumbrances Feinberg
believes the ideal approach could avoid. But we know this
already. The IOM's quite sensible pragmatism seems to distance
policy from fundamental moral concerns and approaches. Indeed,
it is an ongoing struggle in bioethics and policy to straddle these
two apparently incommensurable worlds, the ideal and the practical. In this chapter, I use examples from the work of the National
Bioethics Advisory Commission, of which I was Executive

Director, to illustrate how these approaches can be reconciled. I focus especially on some of the unique features of the public construction of bioethics policy when undertaken by a public advisory body. Of particular importance is the focus on bioethics policy analysis when it occurs *in public*, i.e., where topics are discussed and worked through in public settings (as opposed to in private), and where the public can observe the process from start to finish. In this respect, the *process* is as much the case as the topic or problem being discussed. By "public construction" I am not referring to the process of engaging the public in health policy decision making, about which a helpful literature exists *(5,6)*, but rather to the process of conducting an analysis of a problem by a group authorized by the government to engage in this analysis who are also empowered to provide recommendations for public policy.

## Bioethics Commissions and Public Policy

In more than thirty countries, national bioethics commissions exist to review bioethics issues and to provide advice to their governments *(7–10)*. Committees exist on all the inhabited continents of the globe, in both developed and developing countries, many of which are the first source of consultation for their respective governments on emerging issues in science and technology. Although their actions and authorities differ, these committees share a similar objective: to provide advice and make sound recommendations about issues in bioethics and biotechnology. Relatively little attention has been directed to how bioethics commissions and other advisory bodies undertake the deliberative process, i.e., how they balance or weigh competing values or principles; what presumptions underlie their thinking, and how accurately these *public* commissions reflect *public* thinking. But there is growing anecdotal evidence that these commissions make, and are seen as making, important contributions to public policy.

## US Bioethics Commissions

There is a political tradition in the United States of using bioethics commissions to advise on public policy matters. In many ways, the history of US bioethics commissions parallels the development of US bioethics policy *(3,7)*. Since 1974, there have been six bioethics commissions: the National Commission for the Protection of Human Subjects of Biomedical and Behavioral Research (1974–1978); the President's Commission for the Study of Ethical Problems in Medicine, and Biomedical and Behavioral Research (1980–1983); the Ethics Advisory Board (1978–1980); the Biomedical Ethics Advisory Committee (1988–1989); the National Bioethics Advisory Commission (1995–2001); and most recently, the President's Council on Bioethics (2002–). Other distinguished bodies are entitled to be recognized as "national bioethics committees," including the Advisory Committee on Human Radiation Experiments (ACHRE) (1994–1995) *(11)*; the Fetal Tissue Transplantation Research Panel (1988); the Human Embryo Research Panel (1994); and the National Institutes of Health (NIH) Recombinant DNA Advisory Committee (RAC), even though neither were established as national advisory commissions *per se*.

Some of these bodies have played an influential role in US bioethics policy *(12)*. For example, the National Commission's *Belmont Report* provided the philosophical foundation for the current system of federal regulations for the protection of human subjects *(13)*. The existing system of Institutional Review Boards (numbering between 3000 and 5000) initiated to review research protocols involving human subjects, is one of the outcomes of the National Commission's work.

Similarly, the President's Commission produced an important set of reports, the recommendations from some having made their way into regulations, guidelines, and—in the case of their deliberations on the issue of end-of-life care *(14)*—into a model statute for the determination of death, that has been adopted in all US states. The President's Commission can also take credit for its recommendation that there be a uniform set of human research regulations established in the US. The existence of the *Federal Policy for the Protection of Human Subjects*, also known as the Common Rule because sixteen federal agencies agreed to follow the DHHS regulations, owes its existence, at least in part, to the recommendations in the President's Commission's *Second Biennial Report (15)*.

## The National Bioethics Advisory Commission

The National Bioethics Advisory Commission (NBAC) was established by Executive Order 12975, signed by President Clinton in October 1995. The Executive Order expired on October 3, 2001 and with it, so too did the commission having completed six projects in five years *(16–25)*. NBAC was responsible for advising the National Science and Technology Council—chaired by the President—and other government agencies, on the appropriateness of governmental policies and regulations relating to bioethical issues arising from research on human biology and behavior; as well as the clinical applications of that research. Its eighteen members represented many disciplines; three were designated to represent the general public.

Like other bioethics commissions that have come before it, NBAC adhered to a common procedural requirement for ensuring an open process. All federal advisory committees are required to comply with the *Federal Advisory Committee Act* (FACA), federal law established during the Nixon administration intended to assure the public that groups established by the government to advise it should be accountable. The implementing requirements of FACA are designed to prevent special interest control of the bodies providing advice to the president and Executive agencies, as well as to open public scrutiny to the deliberations of the bodies formulating the advice.

One can think of few examples elsewhere in government where the public construction of policy is so visible. According to Stephen Toulmin, the National Commission's approach to addressing questions involved a nine-step approach including commission deliberation, commissioned papers on analytic and empirical issues, public hearings, site visits, drafting (and voting on) proposed recommendations, before finally submitting recommendations (and their subsequent justifications) to the then-Secretary of Health, Education, and Welfare *(26)*.

NBAC used similar procedures and also sought input from the public as its reports developed, including: making meetings more accessible, setting time aside for public testimony, making extensive use of the World Wide Web, and seeking public comments on draft reports *(27)*. In all, the commission met forty-eight times, usually in day-and-a-half sessions, an average of almost ten meetings per year. The commission usually met in the Washington DC area, but occasionally met in cities around the United States.

## Three Types of Problems Facing Bioethics Commissions

Bioethics commissions in the United States must contend with issues and problems similar to other deliberative bodies engaged in policy analysis. Three common issues are: identifying the problem to be studied; identifying the methods for studying the problem; and identifying the process by which agreement on a problem is reached. FACA ensures that these commissions also share an additional feature, one that distinguishes them from other private deliberative bodies (such as coporate boards, professional bodies, and even academic researchers): that their work is carried out in the sunshine.

### *Identifying the Problem to be Studied*

The Executive Order responsible for creating NBAC provided the process by which topics for reports would be selected:

> In addition to requests for advice and recommendations from the National Science and Technology Council, NBAC may also accept suggestions for consideration from both the Congress and the public. NBAC may also identify other bioethical issues for the purpose of providing advice and recommendations, subject to the approval of the National Science and Technology Council.

In addition, the Executive Order identified the commission's first set of topics. "As a first priority, NBAC shall direct its attention to consideration of: protection of the rights and welfare of human research subjects; and issues in the management and use of genetic information, including but not limited to, human gene patenting." Other issues could be addressed by NBAC if they met the following criteria: (1) the public health or public policy urgency of the bioethical issue; (2) the relation of the bioethical issue to the goals for Federal investment in science and technology; (3) the absence of another entity able to deliberate appropriately on the bioethical issue; and (4) the extent of interest in the issue within the federal government. Its six reports meet these criteria, but their origins are not identical. Three reports—research involving persons with mental disorders *(18)*, research involving human biological materials *(19)*, and ethical issues in international research *(23)*—were selected by the commissioners for study because the issues were timely and tended to satisfy the four criteria just outlined. Indeed, the selection of the international report was due, in part, to public input at some of the first NBAC meetings *(28)*.

But the President specifically requested two reports—one on cloning *(16);* and the other on the use of human embryonic stem cells in research *(20)*. Both requests came with the problem framed. Shortly after the announcement that Dolly had been cloned, President Clinton asked the commission to prepare a report on the ethical and scientific issues, and to submit it within

ninety days. The President's request was clear: to provide "recommendations on possible federal actions to prevent...abuse."
With this framework, the commission limited itself to concerns
regarding the permissibility of public and private involvement in
reproductive cloning. The commission's principal conclusions
and recommendations were as follows:

1.  [That, at that time] it was morally unacceptable for any-
    one in the public or private sector, whether in a research
    or clinical setting, to attempt to create a child by somatic
    cell nuclear transfer cloning.
2.  There should be a continuation of the current moratorium
    on the use of federal funding in support of any attempt to
    create a child by somatic cell nuclear transfer cloning;
3.  Those in the private and nonfederally funded sectors
    should comply voluntarily with the intent of the federal
    moratorium.
4.  Professional and scientific societies should make it clear
    that any attempt to create a child by somatic cell nuclear
    transfer and implantation into a woman's body would be
    an irresponsible, unethical, and unprofessional act.
5.  Federal legislation should be enacted to prohibit anyone
    from attempting, whether in a research or clinical setting,
    to create a child through somatic cell nuclear transfer
    cloning, but that this prohibition should have a sunset pro-
    vision.

These recommendations were focused on the permissibility
of using a particular technology, for a specific purpose—somatic
cell nuclear transfer to create a child—and did not make recommendations on other uses of cloning technology.

The second example of a problem/topic being identified for
NBAC came during another momentous week in the history of
science, when teams led by James Thomson *(29)* and John
Gearhart *(30)* reported in the peer-reviewed literature on
November 6 and November 10, 1998, respectively, that they had

isolated and cultured human stem cells: Thomson, using an embryo donated by a couple who no longer needed it for fertility treatment, and Gearhart, using cadaveric fetal tissue donated following an elective abortion.

Discoveries in this area followed a predictable path of scientific progress since the successful isolation of stem cells in non-human animals in 1981 *(31)*. Thomson and Gearhart's work alone might have been enough to generate a sustained public discussion about the ethical issues associated with this research (coming as they did within days of each other). But when *The New York Times* reported on November 12, 1998, that Advanced Cell Technology (ACT) Inc. of Worcester, Massachusetts claimed that its scientists had for the first time made human cells revert to their primordial, embryonic state from which all other cells develop, by fusing them with a cow egg and creating a hybrid cell, the issue was thrust into the public spotlight.

Jose Cibelli performed the ACT work with human cells in 1996. Using fifty-two of his own cells, some of them white blood cells and others scraped from the inside of his cheek, Cibelli fused each one with a cow egg from which the nucleus containing the DNA had first been removed. From these fifty-two attempts, only one embryo grew and divided five times, generating cells resembling embryonic stem cells. Michael West, the chief executive officer of ACT, was quoted in the *Times* article saying that he was announcing the work in order to test its public acceptability *(32)*. The Biotechnology Industry Organization issued a press release on November 12, 1998, urging the president to ask NBAC to "review all ethical issues...and after reasoned discussion, to make recommendations to the president on how best to assess the implications" *(33)*. Two days later, President Clinton wrote to Harold T. Shapiro, the NBAC chair:

> This week's report of the creation of an embryonic stem cell that is part human and part cow raises the most serious of ethical, medical, and legal concerns. I am deeply troubled

by this news of experiments involving the mingling of human and non-human species. I am therefore requesting that the National Bioethics Advisory Commission consider the implications of this research at your meeting next week, and to report back to me as soon as possible.

I recognize, however, that other kinds of stem cell research raise different ethical issues. ...With this in mind I am also requesting that the Commission undertake a thorough review of the issues associated with such human stem cell research, balancing all ethical and medical considerations. (ref. *20*, p. 89)

The president's two-paragraph letter described the problem to be solved, and so began NBAC's involvement in this issue, the product of which was a one-hundred-eleven-page, thirteen-recommendation report ten months later.

Having a problem selected by means of presidential correspondence does not limit the study or imperil the academic freedom or integrity of the process. On the contrary, given that the mandate of the commission was "to provide advice and make recommendations," it made good practical sense for the specific question being asked to warrant an answer. A more important consequence of this method of selecting the topic was that it placed the commission's deliberations squarely within the public spotlight. On the other hand, interventions from the White House, especially when they came with tight deadlines—e.g., the president requested that NBAC complete the cloning report within ninety days—immediately affected work underway and caused shifts in agenda and other priorities. One consequence in this instance was that NBAC never prepared a report on gene patenting.

## Identifying the Methods of Analysis

We often fail to appreciate that research in public policy, like other areas of scientific inquiry, will succeed or fail based not only on the basis of how well the problem is formulated, but also whether the methods to address the problem are appropriate. Different bioethical problems may benefit from different approaches. Other committees, particularly the Advisory Committee on Human Radiation Experiments (ACHRE) *(11)* were especially adroit at using methods appropriate for the problem they were trying to solve *(34)*. NBAC wisely learned from them and adopted a multi-methodological approach. Each of NBAC's reports were developed by drawing from a set of complementary methods: commissioners would raise and present issues for consideration; staff would prepare briefing materials based on the published academic literature; experts in their respective fields would be commissioned to prepare manuscripts on specific aspects of a problem and present their reports at commission meetings; and public and expert testimony would be sought, and in several cases, a public comment period was established for submission of written comments on draft reports. Elsewhere I discuss the process of public engagement used by NBAC *(27)*.

What is less clear is how the commission articulated the frameworks, principles, and approaches to the various problems it was chose to solve. Each report benefited from scientific, philosophical, jurisprudential, and empirical studies. Several examples are illustrative. The commission's report on international clinical trials *(23)*, took full advantage of a comprehensive empirical study carried out by Nancy Kass and Adnan Hyder of Johns Hopkins University *(35)*. The Kass and Hyder studies, the largest ever carried out on behalf of a US bioethics commission, surveyed more than 500 US and developing-country researchers to understand their experiences with ethical issues in research. These studies were supplemented by qualitative studies carried out by Sugarman *(36)* and Marshall *(37)* in specific locations around the world.

Other empirical studies carried out on behalf of the commission included Elisa Eiseman's inventory of sources of stored tissue in the United States *(38)*, the first such comprehensive look at this issue. NBAC's report on human biological materials *(19)* was directly informed by Eiseman's data, which found that there were approximately 282 million specimens in the nation's laboratories, tissue repositories, and health care institutions—many of which were collected without explicit, informed consent for research purposes.

The commission utilized legal scholarship in several of its reports. In its international report *(23,24)* a comparative analysis was conducted of twenty-three national and international documents addressing the protection of research participants, including guidelines from Brazil, Canada, China, Thailand, Denmark, Finland, India, and Uganda. This analysis illustrated both how far other country's guidelines had come with respect to these issues, and the extent to which US regulations were in need of revision.

Legal analysis informed NBAC's report on embryonic stem cell research. Lori Andrews' analysis of state regulation of stem cell research *(39)*; Brady, Newbury and Girard's analysis of the Food and Drug Administration's statutory authority to regulate embryonic stem cell research *(40)*; Flannery and Javitt's analysis of federal laws, and in particular the Public Health Act, as they pertain to federal funding of stem cell research *(41)*; and Kyle Kinner's informative survey of the history of federal regulation of both fetal tissue and stem cell research *(42)* provided ample legal background for the commission to reference, despite the fact that the principal focus of the report was on the ethical acceptability of federal funding of stem cell research.

Philosophic analysis is abundant in the background materials provided to commissioners. The cloning report benefited from ethical analyses provided by Dan Brock *(43)*. The report on human biological materials *(19)*, used portions of the ethical framework prepared by Allen Buchanan *(44)* , and the stem cell report was very much informed by the thoughtful analyses carried out by John Fletcher *(45)*, Erik Parens *(46)*, and Andrew Siegel *(47)*. Two commissioned papers addressing religious perspectives were prepared for the cloning report *(48,49)*, and the stem cell report devoted an entire volume to the written testimony of ten religious scholars *(50)*. Four of the six NBAC reports contained chapters devoted specifically to ethical issues and several reports made mention, and appreciated the value of, the ethical principles of respect for persons, beneficence, and justice— described in the *Belmont Report*, and more fully developed in many editions of Beauchamp and Childress' *Principles of Biomedical Ethics (51)*. But no report announced at the outset a specific set of "universal laws" or "immutable principles" from which deliberations would ensue, nor did any one ethical theory inform all of the commission's recommendations. And yet no report was hindered by the lack of an articulated ethical framework applied to all situations. Rather, from my vantage point, commissioners rationally and *publicly* deliberated—testing their ideas and perspectives against the views of their fellow members, seeking above all to find consistency and transparency in their conclusions.

An especially valuable example of the challenge of incorporating ethical argumentation into a commission report is Allen Buchanan's commissioned paper for the report on human biological materials *(44)*. Buchanan's ethical framework, recommended to the commission as an approach worth adopting, meticulously described many of the salient interests that weigh in favor of restricted access to, and substantial control by, the source of these materials, and those interests that weigh in favor of fewer restrictions on access to, and less control of, the source of these materials. Like much of Buchanan's work in bioethics, this paper was well argued, tightly reasoned, and compelling. The commission did not, however, *adopt* it as the foundation for the report. This is not to say that the commission's arguments did not reflect their considered judgment of Buchanan's arguments—indeed, some of the justifications for particular recommendations can be traced, indirectly, to the Buchanan presentation.

## Identifying the Process by Which Agreement is Reached

NBAC's stem cell report provides some insight into the issues involved in coming to agreement about morally contested topics in public policy. The first challenge was to satisfy President Clinton's request for an immediate response to his unease about the purported ACT experiment. An already scheduled meeting in Miami, Florida on November 17–18, 1998, provided the opportunity for NBAC to engage this issue. Following a late night drafting session on November 17, and further discussion in open session the next day, a letter was prepared for the president, which read in part:

> The Commission shares your view that this development raises important ethical and potentially controversial issues that need to be considered, including concerns about crossing species boundaries and exercising excessive control over nature, which need further careful discussion. This is especially the case if the product resulting from the fusion of a human cell and the egg from a non-human animal is transferred into a woman's uterus and, in a different manner, if the fusion products are embryos even if no attempt is made to bring them to term. In particular, we believe that any attempt to create a child through the fusion of a human cell and a non-human egg would raise profound ethical concerns and should not be permitted. (ref. *20*, p. 91)

This response paralleled the commission's argument in the cloning report *(52)*, but the more substantive deliberations on human embryonic stem cell research had yet to begin. It was clear from the first discussions of this topic at the meeting in Miami, that most commissioners were prepared to recommend that research using cadaveric fetal tissue from elective abortions should continue to be eligible for federal funding, and that research using embryos remaining after infertility treatments as a source of ES cells should be eligible for federal funding. On the other hand, there appeared to be little support for recommending that research intending to generate embryos solely for research purposes should be eligible for federal funding.

The commission was able to come to this level of agreement because it had found it helpful to frame the problem in terms of the sources of stem cell research: embryonic germ cells from cadaveric fetal tissue, embryonic stem cells from embryos remaining after infertility treatments, and ES cells from embryos created through in vitro fertilization or somatic cell nuclear transfer cloning.

From the outset, NBAC's goal was to develop a set of recommendations that would provide guidance on the appropriateness of permitting the federal government to fund human embryonic stem cell research, and under what constraints. This task was approached with an important principle in mind, namely: "if it is possible to achieve essentially the same legitimate public goals with a policy that does not offend some citizens' sincere moral sensibilities, it would be better to do so" (ref. *20*, p. 57). This principle was daunting to translate into practice. For example, the commission heard testimony from leading academic theologians, including those representing Roman Catholic, Jewish, Islam, Protestant, and Eastern Orthodox traditions—with more than one representative from several of these positions *(50)*.

Many were concerned about the destruction of human embryos that would be the source of stem cells. From this group of scholars there was "thin" agreement about the moral acceptability of stem cell research—i.e., agreement that stem cell research is not inherently immoral and that it has the potential to contribute knowledge leading to therapy, provided that morally legitimate sources of cells are used, and that specific justice and regulatory issues are addressed. Agreement about a principle such as the potential benefit from an area of research, without agreement about the source of stem cells, may not be much agreement at all. The areas of diversity reflected beliefs about the moral status of the embryo, and the reluctance to support any form of cloning embryos for the express purpose of destroying them in research. The commission was faced with an important challenge: on the one hand, it recognized that on issues such as the moral status of the embryo, where such profound disagreement exists, it is unlikely to achieve consensus by sheer force of argument *(20,53)*. On the other hand, the commission recognized its responsibility to provide advice to the President on the matter at hand. It considered several approaches, including one suggested by Alta Charo *(54)* to use political philosophy rather than moral philosophy. In the end, NBAC was informed by an approach first proposed by Patricia King in print *(55)* and in testimony before the commission: "policy in this area should demonstrate respect for all reasonable alternative points of view and that focus, when possible, on the shared fundamental values that these divergent opinions in their own ways, seek to affirm (ref. *20*, p. 51). From this, NBAC constructed what it believed would be a reasonable statement that might reflect areas of agreement:

> Research that involves the destruction of embryos remaining after infertility treatments is permissible when there is good reason to believe that this destruction is necessary to develop cures for life-threatening and/or severely debilitating diseases and when appropriate protections and oversight are in place, in order to prevent abuse. (ref. *20*, p. 52)

Conducting policy analysis in public requires that individuals both express and contribute their points of view as experts, and make calculations about whether (and when) to allow the interests of the group to take precedence. In NBAC's case, consensus was not a goal *at all costs*. That is, the commission did not seek agreement for the sake of agreement alone with compromises negotiated—described earlier by Feinberg as the "need to please voters, to make deals with colleagues." Neither was consensus a requirement, and for good reason: while consensus may be the most expeditious form of substantive agreement on matters of science and ethics policy *(56)*, it is also the weakest form of such agreement subject to breakdown when either certain facts or moral values supporting the arguments change *(57)*.

Unlike the National Commission, NBAC did not take formal votes on its recommendations. Rather, respectful agreement was reached on individual recommendations and on report language for two reasons: first, by virtue of Chairman Shapiro's good leadership and second, by the respect accorded each other's positions. That is not to say that commissioners were always in agreement. Those who found that recommendations, or the arguments supporting them, did not adequately reflect their personal views were encouraged to develop "personal statements" that would be included in the body of the relevant report; this occurred in three reports (ref. *8*, pp. 85–88; ref. *19*, p. 65; ref. *20*, p. 59).

Although each of these personal statements improved the quality and integrity of the reports about which they were written, one statement in particular demonstrated the value to policy analysis of *not* forcing consensus. Recommendation 9 of the report on the use of human biological materials addresses the choices that consent forms might include to allow potential subjects to understand the options available to them, should they be asked to provide samples for future use. The commission recommended that consent forms might include many options, for example:

1. Refusing use of their biological materials to research.
2. Permitting only unidentified or unlinked use of biological materials.
3. Permitting coded or identified use for a particular study only with no re-contact permitted.
4. Permitting coded or identified use of biological samples for a particular study only with further contact permitted to ask for permission to conduct further studies.
5. Permitting coded or identified use of samples for any study relating to the condition for which the sample was originally collected, with further contact permitted to seek permission for other types of studies.

6.  Permitting coded use of materials for any kind of future
    study (ref. *19*, pp. 64,65).

The statements of three commissioners are contained in a foot-
note appearing at the bottom of page 65 of the report, directly
below the body of the recommendation. Three commissioners
(Capron, Miike, and Shapiro) describe their objections to including option 6 in prospective consent forms, and one commissioner
(Capron) describes his opposition to including option 5 in consent forms. These objections, read at the same time as the recommendations themselves, provide readers with a fuller appreciation
of the nuance involved in the recommendation, thus improving
the quality of the analysis.

## Concluding Thoughts: Bioethics in Public

It is simplistic to conceive of policy as having a beginning,
middle, and end. One reason is that the policy products, such as
guidelines, regulations, or legislations are themselves subject to
review, reform, and re-thinking. Nor does policy exist in a
vacuum, unaffected by other policies in different areas. This may
account for why, on July 14, 1999, shortly after concluding its
meeting in Cambridge, Massachusetts (and before NBAC's stem
cell recommendations were finalized), the White House issued a
press release signaling its opposition to NBAC's likely recommendation that research involving the derivation and use of ES
cells should be eligible for federal funding: "No other legal
actions are necessary at this time, because it appears that human
embryonic stem cells will be available from the private sector.
Publicly funded research is permissible under the current Con-
gressional ban on human embryo research."

Seen from this perspective, bioethics policy construction is
more three-dimensional and chaotic than it is two-dimensional
and linear. Bioethics advisory commissions play a part, and only
a part, in the construction of public policy. Bioethics commis-

sions are instruments of public policy analysis, but a commission is only one of many instruments in the orchestra. Moreover, their authority, with rare exceptions, tends to be limited to their advisory role. They do not have the power to implement the recommendations they make, nor to compel government agencies or others to comply with the recommendations.

We are only now beginning to assess the impact of bioethics commissions on public policy. Some work has already been started *(7,11,61)*, but much more is needed.

# References

[1]R.H. Blank and J.C. Merrick, eds. (1996) *Encyclopedia of U.S. Biomedical Policy.* Greenwood Press, Westport, CT.

[2]A.R. Jonsen and S. Toulmin. (1988) *The Abuse of Casuistry: A History of Moral Reasoning.* University of California Press, Berkeley, CA.

[3]Institute of Medicine (IOM). (1991) *Biomedical Politics.* National Academy Press, Washington, DC.

[4]J. Feinberg. (1973) *Social Philosophy.* Prentice Hall, Englewood Cliffs, NJ.

[5]C. Charles and S. DeMaio. (1993) Lay Participation in Health Care Decision Making: a Conceptual Framework. *J. Health Polit. Policy Law* 18:881–904.

[6]B. Jennings. (1993) Health Policy in a New Key: Setting Democratic Priorities. *J. Soc. Issues* 49:169–84

[7]U.S. Congress, Office of Technology Assessment. (1993) *Biomedical Ethics in U.S. Public Policy-Background Paper, OTA-BP-BBS-105.* U.S. Government Printing Office, Washington, DC. June.

[8]National Bioethics Advisory Commission (NBAC). (1998) *1996–1997 Annual Report.* U.S. Government Printing Office, Rockville, MD. June.

[9]S.S. Fluss. (2000) *International Guidelines on Bioethics* [Supplement to *The EFGCP News*]. Council for International Organizations of Medical Sciences (CIOMS), Belgium. Autumn.

[10]J. Glasa, ed. (2000) *Ethics Committees in Central and Eastern Europe: Proceedings of the International Bioethics Conference.* Institute of Medical Ethics and Bioethics Fdn., Bratislava, Slovak Republic.

[11]Advisory Committee on Human Radiation Experiments (ACHRE). (1995) *Final Report.* Oxford University Press, New York, NY.

[12]B.H. Gray. (1995) Bioethics Commissions: What Can We Learn from Past Successes and Failures?, in *Society's Choices: Social and Ethical Decisionmaking in Biomedicine.* (R.E. Bulger, E.M. Bobby and H.E. Fineberg, eds.) Institute of Medicine, Washington, DC, pp. 261–306.

[13]National Commission for the Protection of Human Subjects of Biomedical and Behavioral Research (National Commission). (1979) *Belmont Report: Ethical Principles and Guidelines for the Protection of Human Subjects of Research.* U.S. Government Printing Office, Washington, DC.

[14]President's Commission for the Study of Ethical Problems in Medicine and Biomedical and Behavioral Research (President's Commission). (1983) *Deciding to Forego Life Sustaining Treatment.* U.S. Government Printing Office, Washington DC.

[15]President's Commission. (1983) *Implementing Human Research Regulations: Second Biennial Report.* U.S. Government Printing Office, Washington DC.

[16]NBAC. (1997) *Cloning Human Beings.* U.S. Government Printing Office, Rockville, MD. June. [Note: All NBAC reports are available at http://bioethics.georgetown.edu/NBAC/pubs.html]

[17]NBAC. (1997) *Cloning Human Beings-Volume II, Commissioned Papers.* U.S. Government Printing Office, Rockville, MD. June.

[18]NBAC. (1998) *Research Involving Persons with Mental Disorders that May Affect Decisionmaking Capacity.* U.S. Government Printing Office, Rockville, MD. November.

[19]NBAC. (1999) *Research Involving Human Biological Materials: Ethical Issues and Policy Guidance.* U.S. Government Printing Office, Rockville, MD. August.

[20]NBAC. (1999) *Ethical Issues in Human Stem Cell Research.* U.S. Government Printing Office, Rockville, MD. September.

[21]NBAC. (2000) *Research Involving Human Biological Materials: Ethical Issues and Policy Guidance-Volume II, Commissioned Papers.* U.S. Government Printing Office, Rockville, MD. January.

[22]NBAC. (2000) *Ethical Issues in Human Stem Cell Research-Volume*

*II, Commissioned Papers.* U.S. Government Printing Office, Rockville, MD. January.

[23]NBAC. (2001) *Ethical and Policy Issues in International Research: Clinical Trials in Developing Countries.* U.S. Government Printing Office, Bethesda, MD. April.

[24]NBAC. (2001) *Ethical and Policy Issues in International Research: Clinical Trials in Developing Countries-Volume II, Commissioned Papers.* U.S. Government Printing Office, Bethesda, MD. May.

[25]NBAC. (2001) *Ethical and Policy Issues in Research Involving Human Participants.* U.S. Government Printing Office, Bethesda, MD. October.

[26]S.E. Toulmin. (1987) The National Commission on Human Experimentation: Procedures and Outcomes, in *Scientific Controversies: Case Studies in the Resolution and Closure of Scientific Controversies.* (H.T. Engelhardt, Jr. and A.L. Caplan, eds.) Cambridge University Press, New York, NY, pp. 599–613.

[27]E.M. Meslin. (1999) Engaging the Public in Policy Development: The National Bioethics Advisory Commission Report on Research Involving Persons with Mental Disorders that May Affect Decisionmaking Capacity. *Accountability in Research* 7:227–240.

[28]H.T. Shapiro and E.M. Meslin. (2001) Ethical Issues in the Design and Conduct of Clinical Trials in Developing Countries. *N. Engl. J. Med.* 345:139–141.

[29]J.A. Thomson, J. Itskovitz-Eldor, S.S. Shapiro, et al. (1998) Embryonic Stem Cell Lines Derived from Human Blastocysts. *Science* 282:1145–1147.

[30]M.J. Shamblott, J. Axelman, S. Wang, et al. (1998) Derivation of Pluripotential Stem Cells from Cultured Human Primordial Germ Cells. *Proc. Natl. Acad. Sci. USA* 95:13726–13731.

[31]M.J. Evans and M.H. Kaufmann. (1981) Establishment in Culture of Pluripotential Cells from Mouse Embryos. *Nature* 292:154–156.

[32]N. Wade. (1998) Researchers Claim Embryonic Cell Mix of Human and Cow. *The New York Times,* November 12, A-1.

[33]Biotechnology Industry Organization (BIO). (1998) *BIO Urges President's Bioethics Advisory Panel To Consider Issues Raised By Stem Cell Research* [Press Release]. November 12. Available from: http://www.bio.org/newsroom/newsitem.asp?id=1998_1112_01.

[34]E.M. Meslin. (1996) Adding to the Canon: The Final Report [Book Review of *The Final Report: White House Advisory Committee on Human Radiation Experiments*]. *Hastings Center Report* (Sept/Oct):34–36.

[35]N. Kass and A.A. Hyder. (2001) Attitudes and Experiences of U.S. and Developing Country Investigators Regarding U.S. Human Subjects Regulations, in NBAC, supra reference 24.

[36]J. Sugarman, B. Popkin, J. Fortney, et al. (2001) International Perspectives on Protecting Human Subjects, in NBAC, supra reference 24.

[37]P.A. Marshall. (2001) The Relevance of Culture for Informed Consent in U.S.-Funded International Health Research, in NBAC, supra reference 24.

[38]E. Eiseman. (2000) Stored Tissue Samples: An Inventory of Sources in the United States, in NBAC, supra reference 21.

[39]L. Andrews. (2000) State Regulations of Embryo Stem Cell Research, in NBAC, supra reference 22.

[40]R.P. Brady, M.S. Newberry and V.W. Girard. (2000) The Food and Drug Administration's Statutory and Regulatory Authority to Regulate Human Pluripotent Stem Cells, in NBAC, supra reference 22.

[41]E.J. Flannery and G.H. Javitt. (2000) Analysis of Federal Laws Pertaining to Funding of Human Pluripotent Stem Cell Research, in NBAC, supra reference 22.

[42]K. Kinner. (2000) Bioethical Regulation of Human Fetal Tissue and Embryonic Germ Cellular Material: Legal Survey and Analysis, in NBAC, supra reference 22.

[43]D. Brock. (1997) An Assessment of the Ethical Issues Pro and Con, in NBAC, supra reference 17.

[44]A. Buchanan. (2000) An Ethical Framework for Biological Samples Policy, in NBAC, supra reference 21.

45J.C. Fletcher. (2000) Deliberating Incrementally on Human Pluripotent Stem Cell Research, in NBAC, supra reference 22.

[46]E. Parens. (2000) What Has the President Asked of NBAC? On the Ethics and Politics of Embryonic Stem Cell Research, in NBAC, supra reference 22.

[47]A.W. Siegel. (2000) Locating Convergence: Ethics, Public Policy and Human Stem Cell Research, in NBAC, supra reference 22.

[48]C.S. Campbell. (1997) Religious Perspectives on Human Cloning, in NBAC, supra reference 17.

[49]C.S. Campbell. (2000) Research on Human Tissue: Religious Perspectives, in NBAC, supra reference 21.

[50]NBAC. (2000) *Ethical Issues in Human Stem Cell Research-Volume III, Religious Perspectives.* U.S. Government Printing Office, Rockville, MD. June.

[51]T.L. Beauchamp and J.F. Childress. (2001) *Principles of Biomedical Ethics.* 5th ed. Oxford University Press, New York, NY.

[52]E.M. Meslin. (2000) Of Clones, Stem Cells, and Children: Issues and Challenges in Human Research Ethics. *J. Womens Health Gend. Based Med.* 9:831–841.

[53]T.H. Murray. (1996) *The Worth of the Child.* University of California Press, Berkeley, CA.

[54]R.A. Charo. (1995) The Hunting of the Snark: The Moral Status of Embryos, Right-to-Lifers and Third World Women. *Stanford Law and Policy Review* 6:11–27.

[55]P.A. King. (1997) Embryo Research: The Challenge for Public Policy. *J. Med. Philos.* 22(5):441–455.

[56]T.L. Beauchamp. (1987) Ethical Theory and the Problem of Closure, in *Scientific Controversies: Case Studies in the Resolution and Closure of Scientific Controversies.* (H.T. Engelhardt, Jr. and A.L. Caplan, eds.) Cambridge University Press, New York, NY, pp. 27–48.

[57]J.D. Moreno. (1996) *Deciding Together: Bioethics and Moral Consensus.* Oxford University Press, New York, NY.

# 5

# Finding the Good Behind the Right

## *A Dialogue Between Nursing and Bioethics*

### *Patricia Benner*

> ...There appears to be an internal relation between truth and
> goodness and knowledge. I have argued in this sense from
> cases of art and skill and ordinary work and ordinary moral
> discernment, where we establish truth and reality by an
> insight, which is an exercise of virtue. Perhaps *that* is the
> beginning. *(1)*

Biomedical ethics applied to nursing has been concerned
primarily with eight interrelated areas: (1) clinical competence, judgment, and comportment of practitioners; (2) just allocation of scarce resources; (3) protection of human subjects; (4)
ethical assessment of medical technologies; (5) ensuring patient
rights, including autonomy and informed consent; (6) beneficent
practice; (7) non-maleficence; and (8) social policy related to
health care. These are bold ethical agendas that are still being
worked out and will continue to be central to the ethics of health

From: *The Nature and Prospect of Bioethics: Interdisciplinary Perspectives*
Edited by: F. G. Miller, J. C. Fletcher, and J. M. Humber
© Humana Press Inc., Totowa, NJ

care. In this chapter, I create a dialogue between biomedical ethics, centered on patient's rights, and the internal ethical concerns of the practice of nursing. In the now classic biomedical textbook on a principle-based approach to bioethics, Beauchamp and Childress *(2)*, present four major ethical principles: autonomy, justice, beneficence, and non-maleficence that can be applied in cases of ethical conflicts and dilemmas, to reach ethical decisions. Patient autonomy and informed consent, as overt ethical reforms to paternalism, have exerted a strong moral influence in medicine and nursing. The principle of justice as equal rights of the individual to the public goods of society necessarily focuses on freedom from tyrannies of one group determining the ultimate values or notions of good for any particular individual. As Rawls *(3)* defined deontological ethics, such an ethic requires "thin notions of the good" in order to prevent the imposition of specific notions of good of one group within a pluralistic civil society on another person's notions of the good life. In the wake of the September 11th terrorists' attacks on the American way of life, the experience of the goodness of freedom from infringements on rights and privileges of others becomes fully apparent as a good in itself. This unjust attack on innocent citizens, who became identified as abstract symbols of wealth, capitalism, and freedom demonstrates the importance of "freedom from" the tyrannies of others who would impose their views of God and justice on others. Principles of justice create public spaces where people can meet with well-established ground rules and legal infrastructures to exercise their right to the pursuit of happiness as they define it, not as it is defined or dictated by others. Even though these legal structures are often not more enlightened than the prevailing culture, they have a notable record of defending the private citizen's right to pursue their own happiness and well being, as long as it does not violate the freedom of the same pursuit by others. This so-called negative freedom of rights ultimately cannot be separated from the good experienced in such freedom from tyrannies and coercion.

The focus on autonomy, justice, and individual rights helps institutionalize principles of justice in the decision-making structures and processes of public institutions, where care and meeting of strangers occur *(4)*. As O'Neill points out, inequalities and vulnerability are gathered in public caring institutions, therefore, firm structures and policies that ensure justice are minimally required to prevent neglect and abuse of those who cannot adequately demand or defend their rights. Justice acts as a buffer against prejudices that exclude and deny rights to others. Beneficence and non-maleficence, and a fiduciary relationship are required where rights' bearers cannot understand or demand their rights. However, justice alone is not enough to ensure care of those who are ill, too young, or too old to fend for themselves. The ethical landscape of health care requires a vision of one's basic relatedness to others and notions of a good life in relation to the human condition. For example, in situations of inequality and vulnerability, mercy and generosity will also be required to ensure that rights' bearers who cannot demand their rights will be met in their particularity, protected and nurtured. Sandel *(5)* points out that justice is remedial—it corrects or repudiates injustice. Justice and procedural ethics lodged in institutional policies and processes are necessary, but not sufficient. Moral imagination and solidarity with one's fellow human beings is required to avoid constant infractions of justice.

Nursing, as a women's profession, moved private caring practices into the public domain. Nursing has a complex history in relation to its code of ethics, especially pertaining to the subject of patients and families *(6,7)*. Institutionalizing justice in public caring institutions is fraught with difficulty for many reasons. Care has systematically been relegated to those of lesser power, i.e., primarily women, and the language of care has traditionally been lodged in the private, domestic sphere *(8)*. Institutionalizing justice in the caring work of nurses has been further complicated as a result of the fluctuating relationship of the nurses' institutional work arrangements, between being directly employed by the patient or family, and the hospital or other health care institutions. The role of the nurse in terms of responsibility to the patients and families in public caring institutions has been disrupted by a history of subservience and status inequities between nurses, physicians, and hospital administrators *(9)*. Nurses were taught to be obedient and loyal to hospitals and doctors in order to ensure their safety and continued employment *(7)*. Direct intervention on the patient's behalf relating to good care or ethical consideration, required nurses to be ethical heroes in contrast to ordinary good citizens supported by institutionalized, ethical structures supportive of patient advocacy and caring practices. Despite this tradition of submission to bureaucratic controls, nurses have developed a rich ethical literature about patient advocacy *(10)*. Progress has come with consciousness-raising, associated with the women's movement and the improved education of nurses. However, as the recent shift to a market model of health care delivery has demonstrated, professional nurses continue to have difficulty defending their role and function in public institutions *(11)*. The current International Congress of Nursing Code and the American Nurse's Association Code of Ethics *(12)* clearly articulate that the nurse's first duty and obligations are to the patient.

An ethic of care sometimes exceeds what can be required or demanded by patient rights alone *(13)*. For example, intensive care nurses *(11,14)* frequently tell and write of situations where they experience moral anguish over administering heroic treatments to patients who are receiving futile treatments that prolong their dying because of family requests, or the physician's unrealistic expectations for a cure. In these situations, the anguish of the nurses' felt ethical obligation to stay in the situation, and continue to support the patient by providing as much comfort as possible, while working with the parents or families, and physician to advocate for the patient's best interests, is often described. However, abandoning this situation that they morally disagree with would do an even greater harm to the patient and family. In the case of critically ill neonates, demanding or asserting the infant's rights in the context of parental rights does not adequately meet their felt moral obligation to continue in their supportive and "holding" relationship to the infant and parents, until the parents can face the death of their infant. Nor can they remain silent about the infant's plight *(15)*.

Biomedical ethics has provided an external voice and disciplined thinking about patient rights, and health care professionals' duties and obligations to patients. The need for an external voice aligned with public interests continues to be crucial in the current climate of market models of health-care delivery *(11–13,16,17)*, and as evolving technologies create new moral questions and dilemmas. Health care professionals must not be left to think and decide in isolation about crucial questions concerning rights to treatments, rights to die, informed consent, new biological possibilities in reproduction and fertility, new genetic testing and therapies, cloning of human embryos, and continued threats to equity in health care access. To be more closely aligned with public interests, bioethics will have to grow in its advocacy role as well as in its public policy role in social ethics. Biomedical ethics has given an immense amount of attention to resolving ethical dilemmas in clinical cases. A shift toward advocacy, social ethics, and public policy is particularly important now that access to health care has become an urgent issue for the more than forty million uninsured persons in the United States, and a growing problem for those insured by market-driven managerial systems that control access to health care. It is also an issue of social ethics, that we have increasingly medicalized social problems and issues, so that access to social and caring services are funneled through entry into the medical system *(11,12)*.

Another current challenge for bioethics lies in strengthening and linking the external critique with the moral sources and notions of good *within* the practice of professionals. This is a challenge to create better public language and understanding of the narrative and scientific traditions within particular professions, along with the articulation of notions of good that are central to public-caring practices. We have much to learn about the ethical wisdom, and ethical breakdowns embedded in the day-to-day experiential learning about being a good nurse, physician, or social worker *(18,19)*. Practices are socially embedded and are lodged in narratives and traditions *(20)*. Thus, they cannot be completely objectified or formalized because they are located in the practical world of relationships, action, skillful comportment, and moral agency—as these are lived out in particular situations. Being a good practitioner requires action and reasoning in transition with particular persons in particular situations *(11,21)*. Being-in-relationship to particular persons or situations requires engagement and experiential learning. But since this particular relationship is lodged in a social tradition of schooling, science, and education, those engaging in a practice can recognize obvious instances of excellent or poor practice *(22–24)*. Learning skillful, ethical comportment—central to nursing and medical practice—requires experiential learning that develops skilled know-how, emotional and relational climates, and embodied virtues that are not limited to the required science of the practice. The humanities' side of ethics is required for thinking about good practice in order to provide a continuing renewal of moral imagination within the practice.

The following amended translation of Gadamer by Joseph Dunne (1997), clarifies distinctions between experiential learning and science:

> Experience [Erfahrung] itself can never be science [wissenschaft]. It is in absolute antithesis to knowledge [Wissen] and the kind of instruction that follows from general theoretical or technical knowledge. The truth of experience always contains an orientation towards new experience. That is why a person who is called "experienced" has not only become such through experiences but is also open to new experiences. ...[And] is particularly well equipped to have new experiences and to learn from them. The dialectic of experi-

ence has its own fulfillment not in definitive knowledge, but in that openness to experience that is encouraged by experience itself. (ref. *25*, p.338; translation as amended by Dunne, ref. *26*, p. 306)

Articulating exemplary or excellent practice and the experiential learning of professionals has received less attention from ethicists and philosophers, although there is a growing interest in this narrative approach to ethics *(19,27–30)*. Moral development and experiential learning within excellent practice has its beginning in the work of Aristotle, but has recently been revived as a moral source in health care ethics *(24,31,32)*. Charles Taylor's *(21,33–35)* philosophical writings on moral sources, and his work on public and private life, offer ways to broaden the current bioethical discourse. Taylor argues that people take a stand on their lives through making strong evaluations—some choices are strong, not simple, because they are linked to the person's sense of who they are and what matters to them. Strong evaluations as opposed to simple consumer choices require that a person make qualitative distinctions in ethical comportment and reasoning *(33)*. In nursing and medical practice, ethical (notions of the good and relationship with the other) and clinical discernment are linked. Distinctions between beneficence and maleficence in helping relationships are qualitative distinctions, and are also strong evaluations linked to the well-being of the other, and to fidelity and trust in the relationship *(23,24,36)*.

Aristotle *(36a)* was the first to see the importance of the development of character and moral sensitivities within a practice over time. Joseph Dunne *(26)* notes the following, having mastered the notion of *techne* handed down by Plato:

...[Aristotle] nonetheless stopped short of according to it an unlimited jurisdiction in human affairs. Besides *poiesis*, the activity of producing outcomes, he recognized another type of activity, *praxis*, which is the conduct in public space with others in which a person, without ulterior purpose and with

a view to no object detachable from himself, acts in such a
way as to realize excellences that he has come to appreciate
in his community as constitutive of a worthwhile way of
life [in Taylor's terms, a strong evaluation]. ...Praxis
required for its regulation a kind of knowledge that was
more personal and experiential, more supple and less
formulable, than the knowledge conferred by *techne*. This
practical knowledge (i.e., knowledge fitted to *praxis*)
Aristotle called *"phronesis,"* and in his analysis of it, in
which he distinguished it explicitly from *techne*, he
bequeathed to the tradition a way of viewing the regulation
of practice as something nontechnical but not, however,
nonrational. (ref. *26*, pp. 9,10)

## Articulating a Practice-Based, Relational Ethic

Bioethicists, such as Edmund Pellegrino and David Thomasma
*(37,38)*, have revitalized this discussion of agent-centered
morality and the development of virtues within medicine, in the
Aristotelian tradition of ethical judgment, moral development,
and discernment *(phronesis) (15,39)*. One of the practical
implications of this tradition is that practitioners need to attend
to and articulate what they learn from experience in their everyday practice.

Learning to be a good nurse requires not only technical
expertise but also the ability to form helping relationships and
engage in practical, ethical, and clinical reasoning *(40)*. Six
aspects of skillful, ethical comportment and clinical judgment are
highlighted as central to becoming an excellent practitioner: (1)
linking clinical and ethical reasoning; (2) thinking in action and
reasoning-in-transition; (3) perceptual acuity and the skill of
involvement; (4) skilled know-how; (5) response-based practice;
and (6) moral agency *(11,20)*. In this view, ethical and clinical
reasoning cannot be separated because the vision of what is good,
bad, or harmful dictates sound clinical judgments. The moral
sense of what is good to be and do in a situation guides problem

identification, therapies, and evaluation of care *(11,19,20,41)*. Because nursing is a relatively new discipline and public-caring practice, many of the internal notions of good in nursing are unarticulated, or have difficulty being recognized as legitimate public discourse *(8)*. The limited public language for ethical concerns in nurses' helping relationships provides another strong rationale for articulating the experiential learning and practice-based knowledge of practitioners.

Taking up a practice requires that the practitioner acquire the habits, dispositions, skills, and emotional responses of excellent practitioners. This requires experiential learning. Experiential learning, as Gadamer *(25)* points out, requires having one's preconceptions and expectations turned around, so that understanding, dispositions, and knowledge are changed. The possibilities of moral agency are dependent upon one's vision of a good life, experiential wisdom, skilled know-how, relationship, openness, and responsiveness. In our research, we found that moral agency, as perceived as one's possible impact and influence on the situation for the beginner, consisted in achieving pre-set goals and accomplishing tasks, and being respectful and considerate *(20)*. However, at the proficient and expert levels of practice, moral agency was more attuned to particular concerns in clinical situations and nurse–patient relationships. One's visions of what is possible and capacities to act are based on experiential learning and the skilled know-how to respond and act in particular situations. This stance offers a perspective on the differences between the Kohlberg *(42)* and Gilligan *(43)* visions of moral maturity, and distinctions between a practice- and a principle-based approach to bioethics:

> If one thinks of morality exclusively in terms of *judgments*, which are generated by *principles*, ethics looks like a form of practical reason, and the ability to stand back from the situation so as to ensure reciprocity, and universality becomes a sign of maturity. But if being good means being

able to learn from experience and use what one has learned so as to respond more appropriately to the demands of others in the concrete situation, the highest form of ethical comportment consists in being able to stay involved and to refine one's intuitions. ...Thus when he measures Gilligan's two types of morality—her two voices—against a phenomenology of expertise, the traditional western and male belief in the maturity and superiority of critical detachment is reversed. The highest form of ethical comportment is seen to consist in being able to stay involved and to refine one's intuitions. If, in the name of a cognitivist account of development, one puts ethics and morality on one single developmental scale, the claims of justice, in which one needs to judge that two situations are equivalent so as to be able to apply one's universal principles, looks like regression to a competent understanding of the ethical domain, while the caring response to the unique situation stands out as practical wisdom. If so, the phenomenology of skill and expertise would not be just an academic corrective to Husserl, Piaget, and Habermas. It would be a step toward righting a wrong done to involvement, intuition, and care that traditional philosophy, by passing over skillful coping, has maintained for 2500 years. (ref. *40*, pp. 275,276).

Thinking in action and reasoning in transition refer to practical reasoning or *phronesis* that takes into account changes in the practitioner's understanding of the clinical situation, and transitions in the patient or family condition *(11,19)*. This form of reasoning takes into account changes in perception and directional changes in the patient's condition. Charles Taylor *(21)* has contrasted this form of practical reasoning (which is closer to a moving picture) with scientific and rational-technical reasoning that compares two points in time, by spelling out the situation and the formal criteria for judging the situation into absolute "yes and no" decisions (closer to "snapshot" reasoning). As Taylor *(21,34)* points out, moving through a transition, from a confused or vague understanding to a clearer understanding, is error reducing and clarifies limits and possibilities in the situation. Keeping track of past, changing, and current understandings is the form of practical reasoning or *phronesis* that engages the nurses in their practice. *Phronesis* requires the moral arts of attentiveness and engagement with the other. Threats to this practice are fragmented care-giving episodes, having to delegate too many tasks to other caregivers so that one loses track of the patient's changing condition, and lack of time for the development of relationships. This form of thinking requires more than applying knowledge or subsuming things under categories. As Logstrup points out: "Subsumption is not cognition but an application in which we test whether our cognition was correct" (ref. *44*, pp. 140,141). This insight can be extended to bioethics. Justifying ethical decisions based on ethical principles in bioethics such as autonomy, justice, beneficence, and non-malfeasance does not ensure that the practitioner will notice when these principles are at stake in actual patient-care relationships, or whether the practitioner will be able to develop the relationships that will open possibilities and thinking in action. Ethical principles can enhance accountability and allow for grievance and justification, but the moral agent must breathe life into these principles in action and in fiduciary caring relationships *(45)*. The nurse learns how to be appropriately moved by meeting the other, and by visions of being a good nurse in the particular situation.

# The Role of Emotions in Skillful, Ethical Comportment

Perceptual acuity and the skill of involvement point to the role of emotion and perception in ethical and clinical discernment. Practitioners experientially learn how to be with those who are vulnerable and suffering by doing better or poorer at being emotionally available and attuned to the person's concerns and needs, without being overwhelmed by the other's plight. Learning the skills of involvement (emotional attunement or engagement) teaches the nurse to be with another without becoming either too detached that the needs of the other are not perceived, or too overly identified such that the respect for the identity and separateness of the other are usurped or ignored. Skills of involvement allow the practitioner to have a sense or grasp of their situation, so that a vague uneasiness or sense of disquietude often signals an early, fuzzy recognition of subtle or impending change in the patient. In this view, emotions are schooled to serve rationality and connection. In the early stages of learning to be a nurse, the nurse may sense a generalized anxiety over the patient's vulnerabilities and over his or her own lack of skill and knowledge, but already by the competent stage of skill acquisition, emotions of uneasiness or anxiety have become perceptive, and typically point to a need for attentiveness and puzzle-solving in the specific clinical situation *(11,20)*. Emotional responses become like a moral compass to the excellent practitioner. Nurses who do not experientially learn skills of involvement that allow attentiveness but not over-involvement do not go on to become expert nurses *(20)*. This does not imply that a clinician's sense of salience is infallible, and excellent practice requires that the practitioner stay open to experiential learning and changing relevance to the clinical situation at hand. This social relational space sets up a disclosive space between practitioner and patient or family. Jodi Halpern critiques a detached view of reasoning in medical reasoning, calling for "emotional reasoning," putting forth a revised view of empathy as more than detached cognitive imagination. She states:

> ...Autonomy [interpreted as] non-interference is, in fact, not benign, because the mental freedom to imagine one's own future often comes not from some process inside one's head, but from processes in the social world. It is through emotional communication starting in early infancy that we develop a sense of agency and efficacy, a life-long process.

> When one's identity and goals are stable, a person can be resilient and emotionally independent and withstand social rejection or neglect without being seriously affected. However, when someone's entire sense of self is disrupted, as occurs with suffering and trauma, the impact of not being empathized with can be very severe. (Halpern, 2001, p. 116)

The skillfulness of diagnostic and therapeutic interventions depends upon the nurse's or physician's relationship with the patient in at least three crucial ways:

1. The relationship and the mood or emotional climate of the nurse or physician–patient encounter determines what aspects of the patient's ailments and suffering can or will be disclosed.
2. Knowing the patient in his or her life-world uncovers the contributions and restraints on recovery that a particular person's world makes or could make.
3. The physician's or nurse's caring practices and rhetorical skills determine how and what information the patient will hear from the physician about diagnosis and treatment, and how those may or may not help with the re-integration of the person back into his or her life-world.

All three require *phronesis* and not just *techne (see 46,47)*.

Everyday ethical and clinical comportment are guided, not so much by quandary and extreme cases that fall outside the usual boundaries of good practice, but by usual understandings about worthy, competing goods in particular clinical encounters. For example, the clinician must make qualitative distinctions between care and control, and comfort and suffering; and these distinctions depend on context and relationship *(23)*. Therefore, qualitative distinctions cannot be made through objectification or rational calculation. Emotional attunement creates the possibility of rational action, despite the fact that emotions can also be the seat of irrational actions. Emotional responses can act as a moral compass in responding to the other, and in guiding one's sense of the situation. Emotions, viewed this way, are neither empty of cognitive or moral content nor *necessarily* disruptive and faulty.

The expert nurse can identify or find problems because of perspectives from past clinical situations. Consequently, expert clinicians do not just engage in knowledge utilization; they develop clinical knowledge. A practice, in this view, is not equivalent to matching specific theories to specific situations; it is a dynamic dialogue in which theories and new understandings may be created. The expert is called to think in novel, puzzling, or breakdown situations. Practitioners in particular situations can create what might be imagined to be theoretically implausible actions and outcomes, recombining aspects of theories in novel ways *(48)*.

From a contractual vision of the meeting of autonomous strangers, we do not think of ourselves as being constituted by others, and tend to think of the moral self as that which is "owned" by the self and freely chosen. Care, connectedness, responsiveness, and interdependence are signs of a moral lapse, and are sources of embarrassment for the strictly atomistic vision of the autonomous individual. For the autonomous choice maker, care and caring practices can appear as yet one more set of choices until the position of caring or needing care intrude, because care always implies situated or bounded choice *(34)*. In intimate spheres, loving a child or parent precludes freely choosing to stop caring about the parent or child, though one may physically separate from the other. In less intimate spheres, when one is vulnerable or incapacitated, choices about being cared for and receptivity to care are constrained. Care, publicly and privately, is bound up with the human condition and our commonly experienced vulnerabilities, fears, and dependencies..

Embodied skilled-know is central to thinking in action and excellent ethical comportment on the part of the practitioner *(11)*. Learning how to skillfully respond in the moment of actual concrete situations lies at the heart of becoming a wise and rational nurse. An empathic response is blocked if the nurse does not know how to actively listen, or how to design the best monitoring and dosing of pain medication and comfort measures. Taylor *(49)* and Logstrup *(50)* note that we are in danger of reducing ethical discourse to decisional ethics, so that action, relationship, skilled know-how, and ways of being in the situation are overlooked, while adjudicating right decisions according to well-formulated ethical principles becomes the preoccupation of ethicists.

A number of ethicists have noted that perceptual acuity, i.e., perceiving when an ethical response is required in a situation, and/or discerning what constitutes an ethical challenge, is not automatically ensured by mastering a cannon of ethical principles *(40,51,52)*. Both action and perception are overlooked in ethical theories that focus on rational calculation of rights or consequences of action, rather than on the non-rational aspects of relationship. A broader Aristotelian definition of rational enlarges our moral vision to include action, skillful comportment, relationship, emotional connection, and emotional climates that open up possibilities or close them down. All these practical embodied aspects of the ethical life *(53)* allow people to actualize their notions of good, and even form the conditions of possibility for rational, technical calculation, where formal criteria are established to apply to formal properties of situations *(21)*. Our skillful comportment in a tradition or practice, practical reasoning, and particular relationships are required to initiate and guide any form of rational calculation about ethics.

The claim is that emotions, although nonrational, make rationality possible because emotions are linked to perception and discernment, and make human connection possible *(47,54–56)*. In a practice discipline, emotions are educated to allow for making qualitative distinctions, for attending, for making appraisals of situations, and for creating the possibility of emotional attunement to another's plight and possibilities. The competent performer learns to feel regret over poor performance and satisfaction over good practice, but also learns to relate emotional connections that are facilitating and safe by learning how close or distant, or engaged one must be in certain helping relationships. In psychology this is sometimes called "boundary work," or establishing safe and comfortable boundaries between self and not self. Learning safe boundaries is essential, but the skill of involvement also requires an account of the positive possibilities of opening comfortable, facilitative, emotional climates for growth, recovery, and healing. Practitioners learn the skills of involvement experientially by standing too close, or too far away, impacting the emotional climate and relationship for better or for worse, in order to facilitate the patient and family's well-being *(11,20,46)*.

As Sherman *(55)* points out, this view of the role of emotion is in keeping with Aristotle's views of developing moral character, and of Hume's (1888) view of moral sentiments; but it is stark contrast with Kant's (1785) view of morality as a product of schooling the will in moral principles and obligation. In Kant's view, reason and emotion are radically separated and emotions, such as a predisposition to sympathy, are useful until reason has sufficiently developed to take over the reigns:

> ...duty, or acting on principle, remains the moral motive. That is on theKantian view, as a morally motivated agent, what grounds my reason (for not betraying the patriot to the tyrant for gain, say) is not compassion I happen to feel, but that such action is wrong, and wrong because it manipulates another's rational agency. Emotional motivation is non-moral. On some views, non-moral motivation may be present, but all the same it is not the locus of the morality, or moral worth of the action. (ref. *55*, p. 150)

But in the examples of schooled compassion that Sherman uses *(55)*, and those from studies of nursing practice *(11,19,20,46,57,58)*, emotion enables discernment about when to act and when not to act, and how to be receptive and respectful of the other. Drawing on Taylor *(49)* and Murdoch *(59)*, emotion in the form of loving the good, moves us to act.

Sherman *(55)* points out that Kant is leery of sentimentalism and ineffectual emotion. Logstrup *(44)* takes up the issue of sentimentalism and moralism. Sentimentalism lies in turning emotion in on itself rather than having the emotion direct attention and action to the issue at hand:

> It is characteristic of life-manifestations such as frankness, sympathy, and trust that they divert attention away from themselves and from the individual whose life-manifestations they are, in order to direct attention instead out toward the individual's existence among other human beings and objects. On this ground, life-manifestations bear our existence. As far as frankness is concerned, the person speaking frankly is not aware that he is frank except in situations in which there are costs connected with being frank. Under normal circumstances, he is too absorbed with clearing up some matter. As far as sympathy is concerned, the sympathetic person is not aware that he is sympathetic. He is too absorbed with what ought to be done to relieve the situation for the person who is distressed. As far trust is concerned, the person who is trusting is not aware that he is trusting. He is too absorbed by what is steadfast. In sum, the life-manifestations in question divert too much away from themselves for the person himself to be conscious of them. Which emancipates the contribution of the individual. Absorbed by what should be done—exposing the matter to the light of day, relieving a distressful situation, the person is not aware of life-manifestations. The person identifies himself with a life-manifestation to such a degree that he forgets it. Thus, he is able to focus on his complete concentration upon what the life-manifestation will have for him. (ref. *44*, p. 89)

Sentimentalism for Logstrup, is turning one's attention to the feeling rather than the to issue or task at hand. Sentimentalism takes the form of self-involvement,whereby the person turns the emotion in on itself, or feeds off the risk, vulnerability, or danger of the other's emotional plight. In Logstrup's account, moralism bears a strong resemblance to sentimentalism. A disruptive moralism occurs when the person discovers an emotional impediment or character flaw that blocks effective action and attention. Logstrup gives the example of finding it difficult to arrive at an understanding of a difficult text because of deficient attentiveness and time on the task, so that comprehension is blocked by a superficial approach. Once he has corrected his attention or character problem, then his energy must once again be directed to understanding the text, not to continued attention to his character. Further self-recriminations or censorship from others beyond the person's recognition and self correction, is a form of moralism that prevents action and directedness to the task or concern at hand. Will and emotion are linked rather than oppositional to one another. The schooling of emotion and will require each aspect of the self. "Pure" will and "pure" emotions are a false oppositional account that ignores how will is infused with emotion, and emotions guided by will in the embodied person. The treatment of ethical theories as oppositional to one another, so that one term is defined in terms of what it is not, or one term is defined as if it were a mutually exclusive choice in relation to the other term, is a current problem in the field of bioethics. Kant engages in this kind of oppositional thinking in positing emotion and reason, sentiment, and moral action in oppositional and mutually exclusive ways. This oppositionalism denies the ways that emotions are linked to reason, and the mutual or dialogical form of influence between reason and emotion.

## Grounding Nursing Ethics in Justice and Care

Two landmark feminist writings in ethics, *Justice and Virtue: A Constructive Account of Practical Reasoning* by Nora O'Neill *(4)*, and *Making Virtue a Necessity: Aristotle and Kant on Virtue* by Nancy Sherman *(55)* show the way out of an oppositional stance between a rights based, justice-oriented approach to ethics and a care ethics. Both works are dialogical, and both claim a role for a qualified particularism and for principles. As Sherman notes:

> To make a decision, on the Aristotelian view, is neither to subsume one's choice under some general principle or law, nor to ask whether others could endorse the universalized maxim of one's action. Nor is there the move that others *should* act as we are acting. Thus the orthos logos (right reasoning) of the person of practical wisdom does not involve transforming one's choice into some law-like counterpart, despite a modern bias toward translating the phrase as "right rule." Rather the focus is always on the specifics of the case; wise judgment hits the mean not in the sense that it always aims at moderation, but in the sense that it hits the target for this case. As such, description and narrative of the case are at the heart of moral judgment. This is not to say Aristotle is blind to the fact, so urgent for the Kantian, that we regularly make exceptions for ourselves and that morality must be a matter of confronting squarely those rationalizations. On the contrary, Aristotle insists that the good life is a life studying one's actions, choices and emotional responses, and studying them in a way in which one remains open to criticism and reform [internal critique]. (ref. 55, pp. 244,245)

The practices of nursing and medicine carry within their tradition moral sources for meeting the other in respect and in solidarity with the human condition of embodiment, finitude, vulnerability, and human possibilities. The helping professional must be schooled in skills of involvement, in meeting the other—in receptive ethics *(58)*.

Nurses in practice, even in the most bureaucratic settings, struggle with a relational or care ethic. Nursing as a socially organized set of caring practices brings to the discussion of bioethics, ethical concerns about how to meet, encourage life courage and growth; how to protect, nurture, and comfort those who are vulnerable and in need of care *(46,58)*. But nursing, when truest to its tradition, does this with an acknowledgment of the distinctiveness, and separateness of the other, and with the understanding that the need for care is universal and helpers share the same human possibilities as those they would help *(46)*. This stance is distinct from the technical expert who only holds an external relationship to an object of craft or fabrication. Although nursing as a discipline can claim rationality, knowledge and skillful ethical comportment in its caring practices, it cannot coherently claim, a narrow rational technique that guarantees mastery over the outcomes of caring relationships with concrete, finite others. This places nursing, as a discipline, more firmly in the Aristotelian tradition of *phronesis* and *praxis*, rather than *poiesis* or making, the technical rationality of producing outcomes *(11,19, 20,26)*. A justification of right actions based on moral principles, although useful for institutional policies and procedures, and for justified ethical decision making—especially in dilemma or breakdown cases—is not sufficient for generating or discovering the good in particular concrete, caring relationships. This calls for wanting the other to flourish, to be met and recognized—what Iris Murdoch calls "finding the good in others with no ulterior motive or point to prove" *(1,49)*. That this art would seek the good in situations of risk and vulnerability requires more than a diagnostic armamentarium for fixing pathologies and deficits—it requires that the good possibilities in actual concrete situations and concrete relationships be acknowledged and nurtured. In meeting the other and in caring practices, one finds "situated possibilities" rather than norms or static goals *(46)*. There can be no guarantees in such a fragile and risky set of caring practices. "Helping" that dominates, takes over, or promises what is not feasible, must be vigilantly resisted. The Norwegian nurse philosopher and ethicist, Kari Martinsen *(58)* has written about the necessity of metaphysics and a critical social-ethica for nursing that is life-affirming and nurturing:

Ethics, life-philosophy and metaphysics are cornerstones in nursing. ...Caring for others and loving one's neighbor are the most natural and fundamental aspects of our lives. And also so difficult. *(58)*

Martinsen, influenced by the works of Karl Marx and K. E. Logstrup, critiques the extreme individualism of the modern self-centered and self-assertive individual. In Martinsen's words:

Life with fellow man demands a certain way of living: Not to interfere, control or master the other person. To be open and with feelings for the other yet restrained induces an ethics of reception. The attitude towards life is gratitude. In our world of productivity and results, this becomes critical ethics. ...The battle between conquering and receiving appears in human relations. In human relations, power can be used to destroy the other, or used to expand the other's life-space by receiving him or her. Hope lies in receiving, not in conquering each other. Receiving the other with confidence is criticism of the violent idea of growth, to which we are expected. It is counterweight to progress and competition, which creates loneliness and tension. Receiving the other in confidence is seeing and defending the unqualified human values, in a society which measures them according to qualifications and usefulness. It is seeing what has been given us—seeing the other as creation and irreplaceable. It is seeing the potential in the person who never achieved anything.

Martinsen's work resonates with my work in articulating notions of good in the everyday practice of nurses *(19,46)*. I am convinced that nurses encounter the fundamental demands that the lives of others be received and responded to as members and participants in a common humanity, or as Logstrup and Martisen *(57)* put it, as a response to the fundamental gift of life. First person experience-near stories from nursing practice point to meeting the other in vulnerability, situated possibility, and respect for the life of the other. Nurses have informal narrative dialogues about "knowing the patient and family" *(60)* whereby their judgment is guided by knowing the particular concerns and clinical trajectory of the patient. Within knowing the patient and family, the nurse is able to make qualitative distinctions about what will be experienced as care, and what will be experienced as controlling and dominating *(23)*. Despite our many theoretical and practical reasons for being skeptical about the human possibilities of a receptive ethic, the discovery of these sovereign life expressions against all odds, holds out the distinct possibility of care, and allows us to explore and encounter the sovereignty of the good *(59)*. Charles Taylor makes the point that our sense of moral obligation is dependent on the broader and more fundamental sense of what it is good to be:

> But ethics involves more than what we are obligated to do. It also involves what it is good to be. This is clear when we think of considerations other than those arising from our obligations to others, questions of the good life, and human fulfillment. But this other dimension is there even when we are talking about our obligations to others. The sense that such and such is an action we are obligated by justice to perform cannot be separated from a sense that being just is a good way to be. If we had the first without any hint of the second, we would be dealing with a compulsion, like the neurotic necessity to wash one's hands or to remove stones from the road. A moral obligation comes across as moral because it is part of a broader sense which includes the goodness, perhaps the nobility or admirability, of being someone who lives up to it...

> If we give the full range of ethical meanings their due, we can see that the fullness of ethical life involves not just doing, but also being; and not just these two but also loving (which is short-hand here for being moved by, being inspired by) what is constitutively good. It is a drastic reduction to think that we can capture the moral by focusing

only on obligated action, as though it were of no ethical moment what you are and what you love. These are the essence of the ethical life. (ref. *49*, pp. 14,15)

In recent observations of care planning and reports between hospice nurse and physician, Dr. Derek Kerr explained following the notions of good care with all its particularities and contingencies. For example, rescue treatment is not usually applied in managing terminal pneumonias in hospice care. However, Dr. Kerr explained: "In the case of pulmonary infection which is distressingly malodorous, or provokes a troublesome cough, then oral antibiotics would be prescribed." The person's humanity, facing death, and closing down a life in the context of a particular life and disease, guide medical and nursing interventions *(53)*. The ethical concerns related to patients' rights are different from those related to providing as good a death as possible for a particular patient, once the patient's rights related to dying and treatment are settled. Discernment and risk are ever present as judgments (strong evaluations) are made about increasing narcotics or providing palliative care that will ease the days, and allow persons to face death as they are able to. Notions of good are fragile, and come with risks of not doing or being good in a situation *(61)*.

It is easier to guarantee rights than it is to ensure fidelity to the good in concrete, contingent situations. Practical reasoning *(phronesis)* about facing death, and providing comfort and dignity are not reduced to "choice" or "control" in good hospice care; though choice and control figure into concerns and discernment as the person finds his or her way toward death. Many more particular life goods are at stake than choice: for example, the moral art of holding open a life so that social death does not occur before physical death; so that leave-taking rituals and the human task of facing death are possible. These are the fragile goods that require connection and discernment. They cannot be guaranteed, but they can be nurtured by telling our practice stories where the good is actualized, and by creating work environments that support and encourage caring practices between health care practitioners and patients. Rights are essential and remedial, but not the end of ethical concerns, and they must always be animated by the notions of good that constitute them.

## Notes and References

[1] I. Murdoch. (1992) *Metaphysics As A Guide to Morals.* The Penguin Press, New York.
[2] T. L. Beauchamp and J. F. Childress. (2001) *Principles of biomedical ethics.* (5th ed.). Oxford University, New York.
[3] J. A. Rawls. (1971) *A Theory of Justice. Mass.* Harvard University Press, Cambridge.
[4] O. O'Neill. (1996) *Towards Justice and Virtue. A Constructive Account of Practical Reasoning.* Cambridge University, Cambridge.
[5] M. J. Sandel. (1982) *Liberalism and the Limits of Justice.* Cambridge University, Cambridge.
[6] S. Reverby. (1987) *Ordered to Care: The Dilemma of American Nursing: 1850–1945.* Cambridge University Press, Cambridge.
[7] A. Davis A. and M. A. Aroskar. (1991) *Ethical Dilemmas and Nursing Practice.* Appleton-Century Crofts, New York.
[8] J. C. Tronto. (1993) *Moral Boundaries: A Political Argument for an Ethic of Care.* Rutledge, New York.
[9] R. R. Yarlingand B. J. McElmurry. (1986) The moral foundations of nursing. *Advances in Nursing Science* 8(2): 63–73.
[10] S. Gadow. (1980) Existential advocacy: philosophical foundations of nursing, in *Nursing: Images and Ideals.* (S. F. Spicker and S. Gadow, eds.) Springer, New York, pp. 79–101.
[11] P. Benner, P. Hooper-Kyriakidis,and D. Stannard. (1999) *Clinical Wisdom and Interventions in Critical Care: A Thinking-in-Action Approach.* Saunders, Philadelphia, PA.
[12] M. Fowler and P. Benner. (2001) Implementing the new Code of Ethics for Nurses: An interview with Marsha Fowler." *American Journal of Critical Care* 10(6): 434–437.
[13] P. Benner. (1998) When health care becomes a commodity: the need for compassionate strangers, in *The Changing Face of Health Care.* (J. F. Kilner, R. D. Orr, and J. A. Shelly, eds.) William B.

Eerdmans, Grand Rapids, MI, pp. 119–135..

¹⁴D. A. Asch, J. A. Shea, M. K. Jedrziewski, and C. L. Bosk. The limits of suffering: critical care nurses' views of hospital care at the end of life. *Soc. Sci. Med.* 45: 1661–1668.

¹⁵P. Benner. (1997) A dialogue between virtue ethics and care ethics, in *The influence of Edmund D. Pellegrino's Philosophy of Medicine.* (D. C. Thomasma, ed.) Kluwer Academic Publishers, Boston, MA, pp. 47–61.

¹⁶W. K. Mohr and M. M. Mahon.. (1996) Dirty Hands: The Underside of Marketplace Health Care. *Adv. Nurs. Sci.* 19(1): 28–37.

¹⁷T. Miller. (1998) Center stage on the patient protection agenda: grievance and appeal rights. *Journal Of Law, Medicine and Ethics* 26: 88–99.

¹⁸D. S. Schultz and F. A. Carnevale. Engagement and suffering in responsible care giving: on overcoming maleficence in health care. *Theoretical Medicine* 17: 189–207.

¹⁹P. Benner. (1984) *From Novice to Expert: Excellence and Power in Clinical Nursing Practice.* Addison-Wesley, Menlo Park, CA.

²⁰P. Benner, C. A. Tanner, and C. A. Chesla with contributions by J. Rubin, H. L.Dreyfus, and S. E. Dreyfus. (1996) *Clinical Expertise in Nursing Practice, Caring, Clinical Judgment, and Ethics.* Springer, New York.

²¹C. Taylor. Explanation and practical reason, in *The Quality of Life.* (M. Nussbaum and A. Sen, eds.) Clarendon, Oxford, pp. 208–231.

²²J. Rubin. (1984) *Too Much of Nothing: The Self and Salvation in Kierkegaard's Thought.* (Unpublished doctoral dissertation), University of California, Berkeley, CA.

²³J. Rubin. Impediments to the development of clinical knowledge and ethical judgment in critical care nursing, in Expertise In *Nursing Practice: Caring, Clinical Judgment and Ethics.* (P. Benner, C.A. Tanner, and C. A. Chesla, eds.) Springer, New York, pp. 170–192.

²⁴A. MacIntyre. (1981) *After Virtue: A Study in Moral Theory.* University of Notre Dame, Notre Dame, IN.

²⁵H. Gadamer. (1975) *Truth and Method.* (G. Barden and J. Cumming, Trans.) Seabury, New York.

²⁶J. Dunne. (1993) *Back to The Rough Ground, Practical Judgment and The Lure of Technique.* Indiana University Press, Notre Dame, IN.

²⁷S. Hauerwas and D. Burrell. (1977) From system to story: an alterna-

tive pattern for rationality in ethics, in *Truthfulness and Tragedy: Further Investigations in Christian Ethics.* (S. Hauerwas, ed.) University of Notre Dame Press, Notre Dame, pp. 15–39.

[28]M. Goldberg. (1982) *Theology and Narrative: A Critical Introduction.* Abington Press, Nashville.

[29]K. M. Hunter. (1991) Doctors' Stories: The Narrative Structure of Medical Knowledge. Princeton University Press, Princeton, NJ.

[30]A. Frank. (1998) Just listening: narrative and deep illness. *Families, Systems and Health* 16(3): 197–212.

[31]E. D. Pellegrino. (1979) Toward a reconstruction of medical morality: the primacy of the act of profession and the fact of illness. *J. Med Philos.* 4: 2–56.

[32]E. D. Pellegrino and D. C. Thomasma. (1993) *The Virtues in Medical Practice.* Oxford University Press, New York.

[33]C. Taylor. (1985) *Human Agency and Language: Philosophical Papers, Vol. I.* Cambridge University Press, Cambridge.

[34]C. Taylor. (1989) *Sources of the Self: The Making of the Modern Identity.* Harvard University Press, Cambridge, MA.

[35]C. Taylor. (1994) Philosophical reflections on caring practices, in *The Crisis of Care: Affirming and Restoring Caring Practices in the Helping Professions.* (S. S. Phillips and P. Benner, eds.) Georgetown University Press, Washington, DC, pp. 174–187.

[36]H. L. Dreyfus and S. E. Dreyfus. (1986) Mind Over Machine: The Power of Human Intuition and Expertise in the Era of the Computer. Free Press, New York.

[36a]Aristotle. *Nicomachean Ethics.* (T. Irwin, Transl., 1985) Hackett, Indianapolis, IN.

[37]E. D. Pellegrino and D. C. Thomasma. (1988) *For The Patient's Good: Toward The Restoration of Beneficence in Health Care.* Oxford University Press, New York.

[38]D. C. Thomasma. (1997) The Influence of Edmund D. Pellegrino's Philosophy of Medicine. Kluwer Academic Publishers, Boston.

[39]E. D. Pellegrino. (1979) *Humanism and The Physician.* University Of Tennessee Press, Knoxville.

[40]H. L. Dreyfus, S. E. Dreyfus and P. Benner. (1996) Implications of the phenomenology of expertise for teaching and learning everyday skillful ethical comportmentin, Expertise in *Nursing Practice, Caring, Clinical Judgment and Ethics.* (P. Benner, C. A. Tanner,

and C. A. Chesla, C.A, eds.) Springer, New York, pp. 258–279.

[41]P. Benner and J. Wrubel. (1982) Clinical knowledge development: the value of perceptual awareness. *Nurse Educator* 7: 11–17.

[42]L. Kohlberg. (1981) *The Philosophy of Moral Development: Moral Stages and The Ideal of Justice. Essays on Moral Development, Vol. I.* Harper & Row, San Francisco, CA.

[43]C. Gilligan. (1982) In *A Different Voice: Psychological Theory and Women's Development.* Harvard University Press, Cambridge, MA.

[44]K. E. Logstrup. (1995) *Metaphysics, Vol. I.* (Trans. R.L. Dees, With Introduction.) Marquette University Press, Milwaukee, WI, pp. 140–141.

[45]V. A.Sharpe. (1998) "Why 'Do no harm?" in *The Influence of Edmund D. Pellegrino's Philosophy of Medicine.* (D. Thomasma, ed.) Kluwer, Dordrecht, pp. 197–215.

[46]P. Benner and J. Wrubel. (1989) in: T*he Primacy of Caring: Stress and Coping in Health and Illness.* Addison-Wesley, Menlo Park, CA.

[47]Jodi Halpern. (2000) From Detached Concern to Empathy: Humanizing Medicine. Oxford.

[48]C. Taylor. (1997) Leading a life, in *Incommensurability, Incomparability, and Practical Reasoning.* (R. Chang, ed.) Harvard University Press, Cambridge, MA, pp. 170–183.

[49]C. Taylor. (1996) Iris Murdoch and moral philosophy, in *Iris Murdoch and The Search for Human Goodness.* (M. Antonaccio and W. Schweiker, eds.) University of Chicago Press, Chicago, pp. 3–28.

[50]K. E. Logstrup. (1997) The Ethical Demand. (With Introduction by A. MacIntyre and H. Fink.) University of Notre Dame Press, Notre Dame.

[51]A. J. Vetleson. (1994) *Perception, Empathy and Judgment: An Inquiry into the Preconditions of Moral Performance.* Pennsylvania State University, University Park, PA.

[52]L. Blum. (1994) *Moral Perception and Particularity. Cambridge University, Cambridge.*

[53]P. Benner. (2001) Death as a human passage: compassionate care for those dying in critical care units. *American Journal of Critical Care* 10(5): 355–359.

[54]A. Damasio. (1994) *Descartes Error: Emotion, Reason and The Human Brain.* Putnam, New York.

[55]N. Sherman. (1997) *Making a Necessity of Virtue, Aristotle and Kant*

*on Virtue*. Cambridge University Press, Cambridge.

[56]P. Benner. (2000) The roles of embodiment, emotion and lifeworld in nursing practice. *Journal of Nursing Philosophy* 1(1): 1–15.

[57]K. Martinsen, K. (1989) Omsorg, Sykepleie Og Medisin. Historisk-Filosofiske Essays. [Caring, Nursing and Medicine Historical-Philosophical Essays]. Tano publishers, Oslo, Norway.

[58]K. Martinsen. From Marx to Logstrup. (Trans.) Tano, Oslo, Norway (In Progress.)

[59]I. Murdoch. (1970) *The sovereignty of The Good*. Routledge and Kegan Paul, London.

[60]C. A. Tanner, P. Benner, C. Chelsea, and D. R. Gordon (1993) The phenomenology of knowing a patient. *Image* 25(4):273–280.

[61]M. C. Nussbaum (1986) *The fragility of goodness: Luck and ethics in Greek tragedy and philosophy*. Cambridge and New York: Cambridge University Press.

# 6

## Medical Ethics

### Literature, Literary Studies, and the Question of Interdisciplinarity

#### Kathryn Montgomery

Literature has always been an important part of the discourse of medical ethics. Because short stories, novels, poems, plays, autobiographies, and films offer vivid representations of illness, disability, and dying, they make powerful and effective exemplars in ethics education. In the 1970s, for example, plays like Brian Clark's *Whose Life Is It Anyway* and Peter Nichol's *Joe Egg* responded to the same concerns that prompted early debate in medical ethics and, like that debate, focused attention on the side effects of medicine's technological marvels. A decade later, Larry Kramer's *The Normal Heart* and William Hoffman's *As Is* were the first to pound home the evils of the social refusal to attend to the growing AIDS epidemic. Since the 1970s, autobiographies of illness, like Audre Lorde's *The Cancer Journals*, Anatole Broyard's *Intoxicated by My Illness*, and Reynolds Price's *A Whole New Life*—to say nothing of scores of others by

From: *The Nature and Prospect of Bioethics: Interdisciplinary Perspectives*
Edited by: F. G. Miller, J. C. Fletcher, and J. M. Humber
© Humana Press Inc., Totowa, NJ

people unknown before their diagnosis—have found a ready audience *(1)*. Physician-authors, like Richard Selzer, John Stone, Perri Klass, and more recently Rafael Campo and Michael Stein, have revealed the texture of medical practice. Others, like Donald Hall, Lorrie Moore, and Deborah Hoffmann, have made riveting poems or stories or films about the illness of family members *(2)*.

Fiction, poetry, drama, and autobiographical essays take up these issues not because they are central to medical ethics but because illness and death are part of the human condition that imaginative writing exists to explore. This is not a recent phenomenon: *Philoctetes* and *King Lear* are as relevant to contemporary moral discourse as *ER*. But these days, with many human ills caught in the prolonged embrace of what Lewis Thomas called "half-way technology," medicine has become central to the way we think about the question of meaning in our lives *(3)*. Much of contemporary literature concerns not just illness but its medical treatment, the moral choices that treatment engages, and the failures of human compassion that too frequently accompany our trust in technologized care.

Literature's contribution to discourse about values and behavior, nevertheless, can be easily overlooked. Drama, poetry, and fiction make no explicit argument; stories seem to be always just there. Irony, revelation, and meaning itself all depend on what the observer already knows about human beings, the acts they are likely to commit, and the justification they offer for them—and much of this knowledge has been learned in turn from stories. We do not ordinarily reflect on this: stories are simply our element. We ignore the fact that moral lessons are conveyed in gossip, neighborhood rumors, and office politics. We forget that once we were children hungry to hear stories that made sense of the world. We take for granted the movies, television, and fiction that enable us to look into the abyss—or soothe us past the temptation to peer in. So, too, has it been for literature in medical ethics. This essay traces the growth of literature's use in medical education and the studied distance the field of literature and medicine at first kept from ethics. It describes narrative ethics and the arguments in defense of narrative's use in medical ethics; it puzzles over questions of the mutual influence of literature, literary theory, ethics, and medicine, and concludes with the narratological critique of medical ethics and the prospects for a genuine interdisciplinarity.

## Education for Medical Practice

Literature was widely used in the 1970s by the early teachers of medical ethics. In fact, half the contributors to the first volume of *Literature and Medicine* (1982) were physicians and religious studies scholars testifying to the efficacy and value of literature in the medical-ethics classroom. Literature, which for them meant realistic fiction, illustrated the moral problems that confronted physicians and patients. These early teachers looked for stories that illustrated the questions that at the time characterized medical ethics: should the dying be told the truth? Does mortal danger override a patient's refusal of treatment? Who in a crazy world is insane? The texts they assigned have since become the canonical texts of literature and medicine: Leo Tolstoy's *The Death of Ivan Ilych*, William Carlos Williams' "The Use of Force," and Anton Chekhov's "Ward No. 6."

Literature scholars, slowly added to the medical faculties, increased the canon but did not alter the rationale much at first. Literature, surprisingly, fit the aims and methods of medicine, especially clinical education, quite well. Unlike the social sciences—but very like medical practice—literature is concerned with accounts of human behavior and values as manifested in the individual and not in the aggregate. Like the physician, the writer focuses on details, alert for the telling oddity. Such clues not only categorize the present instance and render it enough like earlier experience to be understood, but also mark its uniqueness and potential narratability. "Happy families," Tolstoy observes at the beginning of *Anna Karenina*, "are all alike. Each unhappy family is unhappy in its own way." He could have said the same of health and illness.

However, it would be years before the interpretive drive shared by medicine and literature or their common narrativity would license the medical-school use of any literary text whatever: Homer or Henry James, Dante or Dostoevsky. Instead, the rationale for including literature in a medical curriculum, its usefulness to medicine, at first meant stories about illness or the patient–physician encounter. There is a sound pedagogical reason for this. In medical textbooks, patients are represented as symptoms, syndromes, body parts, test results. There are no whole people who are ill, nor any physicians, nurses, or families of the ill or dying *(4)*. For an understanding of illness or doctoring or tending to the needs of a sick person, literature and the theater, including television, are medicine's best resources. In medical humanities courses, the list of teachable texts soon expanded to include stories and poems by physician-writers and accounts of the old, the poor, the disabled, the female, ethnic minorities—all those potential patients with whose experience a medical student might not be familiar.

Literature, in the view of these early medical humanists, was a reliably accurate representation of social and physical reality, and its function was to provide its readers a broader education through vicarious experience. For physicians and medical students, that meant illustrative, richly detailed case studies. Far from offering a challenge to medical ethicists' narrow construals of principlism—what James Childress and Tom Beauchamp have defensively labeled "deductivism" *(5)*—scholars in the field of literature and medicine only indirectly claimed literature's relevance to medical ethics. How it might be relevant beyond its "handmaid" function—supplying illustrations of moral problems—was seldom discussed. Literature's obvious conjuncture of culture and psychology, fields of study then widely believed in the social sciences to be radically separate, or its representation of non-dilemmatic everyday experience—or how these matters might be useful in medical ethics—received no attention. Joanne Trautmann Banks' *Healing Arts in Dialogue* (published a number of years after its seminal discussions took place) was an exception; but its reported conversations among physicians and teachers of literature focused, not on medical ethics, but more generally on literature's practical and pedagogical contributions to medical education *(6)*. This became the pattern. Early essays in literature and medicine executed an end-run around the question of literature's relevance to ethics. Focusing instead on its relevance to medicine, they called attention to the conversation about values opened up by reading literature in the medical school and the hospital without directly challenging the prevailing narrow assumptions about what constituted ethical discourse in medicine.

It was physicians and theologians, once again, who pointed out literature's relevance to medical ethics, and they used it to challenge the status quo. Dissatisfied with the dilemmatic or quandary-based character of medical ethics *(7)*, they called attention to the larger relationship between literature and the moral life. Indeed, for more than a decade, the most nearly theoretical document in the field of literature and medicine remained Robert Coles' (1979) *New England Journal of Medicine* article, "Medical Ethics and Living a Life." He distinguishes between the philosophical analysis of particular issues in medical ethics and moral reflection on the larger question of how life is to be lived; reflection available to everyone, he points out, through reading novels *(8)*. Anne Hudson Jones, in her 1987 essay, "Literary Values: The Lesson of Medical Ethics," names four other scholars, who quite early on used literature or literary theory to challenge medical ethics' unalloyed reliance on analytic philosophy: Stanley Hauerwas and David Burrell, Larry Churchill, and Warren Reich *(9)*. There is not a literary scholar among them. Instead, they were narrativists who, like Alasdair MacIntyre, saw moral life as a sustained practice, knowable principally through a narrative of the circumstances and decisions of that life *(10)*.

Where were the literary scholars? For the most part they avoided direct engagement with ethical questions. There was a reason for this reticence. The formalist New Criticism that had shaped the education of most of the early literature scholars in the medical humanities had condemned attention to the author's intention and the audience's response as serious critical fallacies. The author's purpose, the effect of the text on the reader, and the more general matter of literature's participation in moral discourse, were all viewed as irrelevant to the work of art and to a hard-won modernity that held the work to be sufficient unto itself. At their best—as in the work of Lionel Trilling or F. R. Leavis—these "fallacies" were the legacy of Matthew Arnold, who believed that literature might substitute for a dying religion. At their worst, appeals to the author's intention and the audience's response were lapses into Victorian moralizing that had led as late as the mid-1960s to such embarrassments as governmentally banned books and censored editions of the classics *(11)*.

But pure aestheticism is a frail justification for literature's presence in the curriculum, in medicine or out. The distance between the principles of New Criticism and the unacknowledged justification of classroom practice was immense. Thus, when literature was added to the medical curriculum with the apparently straightforward, simple Horatian rationale that literature teaches as well as delights *(12)*, the few literary scholars in medicine did not challenge it. The field was too new and tenuous, and Horace's claim was literature's best justification with medical school deans and clinical colleagues. This is less duplicitous than it seems. The complexities of representation and the problems of interpretation inherent in the act of reading are inalienable aspects of discussing literature in an open classroom. Does the reader read what the writer writes? How can we know? Because every literary text is complex, potentially ambiguous, and inevitably situated; interpretation is the "basic science" of literature and literary theory. Epistemological questions thus have had a kind of stealth existence beneath the radar of literary–theoretical correctness. Marxist or formalist-New Critical, feminist or deconstructionist, the reader must make reliable (if multiple) sense of the text. Early teachers of literature in medicine could use literature's relevance to medical practice and to questions of a good life in medicine as selection criteria—even proclaim literature's representational virtues to curriculum committees—without entirely giving up the New Critical standards that designated the texts they chose as powerful works of art. But the price was silence on important critical questions. With regard to every sort of philosophical or theoretical question, then, the early literature and medicine scholars, very much outnumbered, lay low. They focused instead on the fictional representation of illness and doctoring and the usefulness of literature to medicine.

Nor did this theoretical reticence change when mainstream literary studies exploded in a burst of post-structuralist energy: reader response, authorial representation, interpretive reliability, all became vital issues. Scholars in literature and medicine then might have moved beyond their service-oriented, enlightenment claims to argue literature's epistemological complexity or to challenge narrower concepts of representation and interpretation in medicine and medical ethics. But the strengths of literature in medical education—its assumption that literature mirrors the social world and its attention to interpretation of texts—were devalued by post-structuralist discourse. Thus, for its first decade and more, literature and medicine remained less a discipline than a practice of attentive and relevant teaching together with the explication of texts to enrich that teaching. The field grew and flourished in the 1980s without much in the way of justification or theory.

The interest in literary theory with its potential for medical ethics, when it belatedly began, came not from literature's relation to medical ethics but, in a more roundabout way, from its steady and deepening attention to medicine. That, after all, is what paid the bills. Beginning in the mid-1980s, scholars in literature and medicine brought the insights of literary theory, continental philosophy in disguise, to bear on medical texts and, particularly, on those texts that structuralism had taught them to find not on bookshelves but in the world at large. "The Physician as Textual Critic" described the clinician's interpretive task; Rita Charon's "Doctor–Patient, Reader–Writer" described the patient–physician encounter as a text jointly authored by patient and physician *(13)*. The field's first controversy, occasioned by David Barnard's "A Case of ALS," grew out of an attempt to apply literary critical ideas to a case report and brought such basic literary concepts as narrator, genre, and point-of-view to medical cases *(14)*. In a series of articles, Suzanne Poirier explicated the medical record as narrative; I wrote about the narrative epistemology of medicine; and William J. Donnelly used narrative insights from historiography in proposing reform of the case history*(15)*. In extending the methods and assumptions of literary study to non-literary texts, this work covertly widened the scope of ethical inquiry to take in the ordinarily unexamined, everyday elements of medical practice.

The bold claim that literature is itself moral discourse came from philosophers. Beginning in the early 1970s, Stanley Cavell used complex readings of literary works to address the issues of morality and public policy. Iris Murdoch, novelist as well as philosopher, maintained in *The Sovereignty of Good* that literature is fundamental to the education of moral perception. Bernard Williams suggested that literature supplies the meaning that universalizing ethical systems necessarily leave out *(16)*.

These arguments go well beyond viewing literature as a source of information about the human condition or as merely a more vivid form of sociology. They first hypothesize that as human beings we understand our lives narratively and then locate our acquisition of moral knowledge in the act of reading or hearing those narratives. This position in no way denies that human beings abstract from experience, formulate rules, or perceive new situations in light of these abstract categories, but neither does it relegate narrative knowing to second place. Each is a part of human knowing. Studies in cognitive psychology and artificial intelligence support this view. Jerome Bruner, following William James, distinguishes narrative knowledge from hypothetico-deductive knowledge and finds in it the motive for our earliest language acquisition *(17)*. Roger Schank models human learning on machines using narrative scripts, the "pattern recognition" familiar to physicians *(18)*. Literature, more specifically (and not simply narrative), is established as the site of moral development by Martha Nussbaum and Wayne Booth. In *The Fragility of Goodness*, Nussbaum argues that Greek tragedy is an essential part of the philosophical examination of the role of chance in the moral life *(19)*; in *The Company We Keep*, Wayne Booth investigates the character formation (for ill as well as good) that is an unavoidable aspect of reading novels( *20)*.

By the 1990s, ethics had been restored to respectability in mainstream literary studies *(21)*. Even deconstructionists, the formalist hold-outs who believed words to be understood only in reference to other words, texts only in relation to other texts (and then never finally), found it necessary to address the relevance of texts to lives after the discovery of prewar anti-Semitic newspaper columns Paul de Man wrote before emigrating to the United States and creating the deconstructionist school here *(22)*. Works of mainstream literary scholarship as well as those in literature and medicine turned to Hans-Georg Gadamer, Mikhail Bahktin, and Emmanuel Levinas to find a dialogic balance between the aesthetic and the ethical *(23)*.

In the wake of this literary-ethical breakthrough, literature and medicine's reluctance to theorize its relationship to medical ethics and the moral life looks like an opportunity missed. But it was the result, too, of an oddly prescient pragmatism that focused literary attention on medicine instead. By the time literature and medicine had absorbed the return of ethics to critical respectability in mainstream literary studies, its scholars had been challenging for a decade the narrow understanding of knowledge and representation in both clinical medicine and medical ethics without much distinguishing between the two. Medicine provided a model for this meld since it often takes for granted its inherent moral character. Doctoring not only is (or should be) a series of acts undertaken for the good of the patient, but in its reliance on narrative, it guarantees that physicians will, as Jerome Bruner terms it, "go meta," reexamining what they have done and why they do it *(24)*. This blurring of the line between medicine and its morality, a failure to separate clinical practice and ethics, for me at least, has come to be a virtue *(25)*. Theoretical essays in litera- ture and medicine since the late 1980s have continued to follow the earlier pattern, concerning themselves less with literature's contributions to medical ethics than with its use in inculcating clinical skills and competencies that make up the sound, moral practice of medicine *(26)*. Rita Charon and her colleagues set out a compendium of literature's usefulness in "Literature and Medi- cine: Clinical Contributions" *(27)*.These include insights into the lives of patients and the social importance of medicine, a new approach to the medical ethics and perspective on the work and texts of medicine. A subsequent article on literature's place in medical education in an issue of *Academic Medicine* devoted to the medical humanities emphasized literature's contribution both to clinical skills and to physicians' moral life *(28)*. This perspec- tive, that ethics is a part of medical practice and both clinical medicine and medical ethics are interpretive acts, challenges equally the top-down theories of medical ethics and the claims of philosophy to hegemony in that field.

## Narrative Ethics

What then is "narrative ethics"? For almost a decade the term has vexed and annoyed a number of medical ethicists, who have re- garded it, no doubt, as an oxymoron. What can the term mean? Like the description of medicine as that Aristotelian impossibil- ity, a science of individuals, "narrative ethics" designates at once (and contradictorily) an interest in particulars and a more gener- alized and generalizable, analytic view. This doubleness—an attention to the circumstances of a case without rejecting ethical principles—links it to other interpretive, contextual approaches to ethics: casuistry, phenomenology, hermeneutics, and pragmatism *(29)*. Narrative ethics has been variously described. In its simplest non-literary form, it is an attention to the patient's story. This, of course, is the focus of traditional psychotherapy, and for more than two decades, Oliver Sacks' case studies have pointed out the relevance of the patient's experience to neurology, that other specialty in which the self is often indistinguishable from the malady. But much of the rest of medicine can be practiced quite differently. Lip service is given to the importance of the patient's story, but technologized subspecialties and a highly mobile population—to say nothing of the recent commodification of medical care—have diminished the apparent utility of the story even as it has shrunk to a "social history" consisting only of the patient's alcohol use and pack-years of smoking.

Attention to the patient's story, however, remains important both medically and morally. Rita Charon emphasized its centrality to the moral development of clinicians when she asked students who had just finished their first patient work-ups to imagine (and write) the stories their patients would tell about their illness *(30)*. Howard Brody in *Stories of Sickness* and Arthur Kleinman in *The Illness Narratives* argued the importance of the patient's experience and the narrative shapes it takes, both for the patient–physician relationship and for healing (and sometimes cure) *(31)*. In the early 1990s, Steven Miles and I experimented with collecting richer narratives of ethically troubling medical situations *(32)*. Our working hypothesis was that the conventions of the medical case genre and the habit of top-down reasoning from principles in medical ethics between them eliminated the contextual details that in real life often matter most. We learned a great deal: "futility" is always defined *in situ*; caretakers frequently need care themselves if they are to make good decisions; ever larger stories are always possible and, a corollary, the "whole story" is a more or less arbitrary construct. Most important of all, ethical decisions are inescapably situated both interpersonally and sociopolitically: in the patient's need, the physician's habit and anxiety, the family's hopes for a future.

Rita Charon formalized the taken-for-granted role of narrative in medical ethics in her deceptively modest 1994 manifesto, "Narrative Contributions to Medical Ethics" *(33)* She names four. Narrative is essential for recognizing an ethical problem, its stakeholders, and the coherence of the patient's life. It is part of the formulation of the case *qua* case and inevitably involves a point of view. It enables the interpretation of the ramifications and ambiguities of the case and the negotiation of its meaning among interested parties; and it participates in the validation of this interpretation as the best available understanding of events, one that can authorize action. A grasp of these mutually modifying, simultaneous activities, Charon argues, constitutes a narrative competence essential to ethical practice. They do not replace ethical principles and guidelines, but neither is there any aspect of the use of principles that is left untouched by narrative.

Several typologies of narrative ethics have been suggested. Stressing its compatibility with clinical thinking, Anne Hudson Jones, in a 1997 *Lancet* essay, locates narrative ethics in the case-based reasoning that characterizes both clinical and moral casuistry, in the rhetorical framework of medical ethics more generally, and in the earlier uses of literature as a representation of the patient and a source of moral reflection *(34)*. Thomas H. Murray examines four possible meanings of narrative in ethics: (1) narrative as moral education; (2) as moral methodology; (3) as a form of moral discourse; and (4) as a part of moral justification *(35)*. Only the first, he believes, is compatible with a conception of ethics as a set of moral propositions: narrative "as merely useful tools for filling in the blanks of previously composed moral syllogisms" (p. 9). The other three, he believes, alter the prevailing concept of "doing ethics" (if not necessarily, I would argue, changing its practice much) and give narrative a substantive place in medical ethics.

Narrative ethics appeals to many, including physicians, who have been troubled by the shallow exercise of principlism, but it also has been found wanting by a number of philosophers in medical ethics. A narrative approach to ethics is not systematic. It has been described as simultaneously discarding principles and failing to challenge prevailing ethical theory *(36)*. It tells us only what we already know; it is open to emotion and is therefore corruptible *(37)*.

The most interesting critiques have come from within the field and from philosophers comfortable outside Anglo-American analytic philosophy. Joanne Trautmann Banks and Anne Hunsaker Hawkins remind us that narrative and narrative theory are not the whole of literature's contribution to ethical discourse. Banks points out the value of drama as a representation of conflict both within and between moral agents; Hawkins describes poetry's access to "epiphanic knowledge," a revelation of meaning and value that may be neglected or ignored in medicine and the ordinary practice of medical ethics *(38)* In addition, David Morris and Arthur Frank have criticized the scope and orientation of narrative ethics. In *Illness and Culture in a Postmodern Age*, Morris faults scholars in literature and medicine for not making a place for emotion in their accounts of narrative and narrative rationality; in *The Wounded Storyteller*, Frank charges narrativists with ignoring real accounts of people who are or have been ill, leaving narrative ethics in thrall to medicine *(39)*.

The best philosophical critiques of narrative ethics are those of John D. Arras and Hilde Lindemann Nelson, philosophers well versed in casuistry and literary theory, respectively, who are able to write perceptively about both the value of narrative ethics and the problems it does not solve. Arras examines three sorts of narrative ethics for their implications for moral justification *(40)*. The first, which, following Stanley Hauerwas and Alasdair MacIntyre, holds that cultural narratives and the individual's role therein are the ground of a moral life, he finds to be implicitly relativist *(41)*. The post-modernist narrative ethics examined in Arthur Frank's work, but traced from Jean-François Lyotard and Richard

Rorty, he finds subjective, even solipsistic, and without the means of forming judgments. But because Rita Charon's "Narrative Contributions" reaffirms the need for ethical principles, he can grant that narrative is "important...to ethics as traditionally conceived":

> Narrative provides us with a rich tapestry of fact, situation, and character on which our moral judgments operate... [Without narrative] the moral critic cannot adequately understand the moral issue she confronts, and any moral judgments she brings to bear on a situation will consequently lack credibility. (pp. 82–83)

He concludes that narrative, as characterized in Charon's approach, is "an essential element in any and all ethical analyses, [one that] constitutes a powerful and necessary corrective to the narrowness and abstractness of some widespread versions of principle and theory-based ethics" (p. 84). "To paraphrase Kant," he summarizes, "ethics without narrative is empty" (p. 83).

Why then, one is forced to wonder, does Arras also describe narrative as a "supplement to (or ingredient of) principle-driven approaches," or a "supplement or handmaid to principles and theory" (p. 68)? How can an essential aspect of the acquisition, use, and understanding of principles be merely a supplement? Perhaps principles are merely shortcuts or a supplement that facilitates the use of narrative. Neither Arras' concluding view of narrative as essential to ethics nor his understanding of Rawls' reflective equilibrium warrants awarding narrative this second-class citizenship *(42)*. These labels appear in his account of Charon's argument as if they might be hers. But he has misread her affirmation of ethical principles. She is not arguing for the inclusion of narrative in medical ethics but describing ethics as it is practiced in medicine as inalienably narrative. Hers is a principled focus on what is, after all, important: not the theory and conduct of medical ethics, but "ethical medicine" *(43)*.

The real question for philosophers examining the role of narrative in medical ethics is what ethics would be *without* narrative. We know the answer: symbolic logic, that powerful and objectifying tool—the basic science of analytic philosophers—that by mid-century had rendered philosophical ethics dry as dust. Its goal is to abstract moral situations into universalizable, logical form, and letters and equations are its signs. This is the ethics that, Stephen Toulmin pointed out, medicine saved the life of *(44)*.

There is no narrative ethics, except as a retronym. Ethics *is* narrative. It also must invoke or refer to rules or principles. Each side in this tug-of-war between the particular and the general, with its very different skills and "basic science," sees the whole as its property. Philosophers seek to objectify, and propositional logic is their instrument of choice. Narrativists particularize, and occasionally, in a stable community or when there is consensus, they are inclined to think the principles will take care of themselves. But medical ethics is a discourse about human action and meaning, the practical application of what is known to what is done in the world. Medical ethics is not abstract (or very abstractable) because its objects exist in time. The discourse of medical ethics has the emblematic shape of an hourglass. Experience in all its complexity pours in at its wide mouth and is sorted narratively, as is our human habit. Analytic argument is invaluable at the narrow center where values and principles in the abstract inform our interpretations. Their real-world meanings—with refinements and exceptions—are worked out beyond that point in an enlarged world of action and conflicting views.

## Literature, Literary Theory, and Medical Ethics: Questions of Influence

If literature and medicine has been the doppelganger of medical ethics, and narrative has been a part of ethics all along, how has the ethics movement affected literature and literary study? The answer at first glance seems to be that it hasn't in the slightest: topics that have engaged both were simply "in the air." Literature's traditional concern with finitude and with the individual's struggle with self or society has found tellable stories in the last twenty-five years in medicalized illness and technologized dying—just as the same concerns have increasingly been the subject of newspaper and television stories, ethical debate, and conversations among family and friends. It would be as difficult to attribute to medical ethics Anna Quindlen's novel about assisted suicide, *One True Thing* (where one suspects it) as Sharon Olds' closely observed poems of a last illness, *The Father* (where one doesn't).

Although it would be a stretch to suppose that medical ethics has encouraged the flowering of autobiography and its near-clinical revelation of more intimate moral landscapes—abuse, dysfunction, alcoholism, rape—ethical questions (or the cultural forces shared with medical ethics) have encouraged the flood of contemporary essays about disease states and medical therapy. And as medical–ethical problems have come to be understood as more nuanced, the audience for the first-person medical essay has grown. Sherwin Nuland has given us fascinating accounts of pathophysiology; Jerome Groupman has written moving accounts of puzzling, challenging patients; and Atul Gawande has reported on the conflicts and anxieties of his surgical education: experimental procedures, mistakes, the virtues of expertise *(45)*. Neither fictional nor entirely autobiographical, these essays take medicine as a human frontier, and molecular biology and the human genome project, as territory for exploration. The operating theater is the site of daily defiance of old human limitations; the workings of the body and attention to the dying are sources of wonder. Just as the flood of autobiographical accounts of medical education and its brutalities in the 1980s coincided with, and may have fueled, concern for the worst absurdities of the standard medical curriculum *(46)*, these personal medical essays parallel an increased interest by medical ethicists in the moral experience of the physician. They rival fiction in supplying the rich description necessary for good ethical decisions about medicine.

The effect of medical ethics on literary theory is harder to specify. Both medical ethics and the study of literature since the 1970s have benefited from a broader, less formalist, more inter-disciplinary orientation in the humanities. Genres have blurred and, with them, disciplinary boundaries *(47)*. Physical sciences now seldom supply the measure of truth in the humanities and the social sciences. Explanations in every field are more global, less mechanistic and paternal. Most important, modes of intellectual and disciplinary representation have themselves become the objects of inquiry. In such an intellectual climate, any claim to dominance, especially philosophy's customary claim to be the discipline of ultimate resort—or even to supply the essential lan-guage for moral discussion *(48)*—is not simply arrogant, but itself open to intellectual scrutiny. It could be that the existence of the medical ethics movement encouraged Martha Nussbaum's argu-ment that the Greeks understood the moral life to be more than isolated rational decisions or Wayne Booth's examination of the moral subtleties at work in the interpretive act of reading. Cer-tainly, each had colleagues who were active early participants in medical ethics. But communal influence merges with the obvious fact that all this work was nourished in the same cultural medium: a general concern with the self, its identity, and its values *(49)*.

Questions about the mutual influence of medical ethics and literature are finally misplaced, however, for medicine itself has been the strongest extra-disciplinary influence on both fields. In a post-modern era, the care of the ill remains a quintessentially modernist activity. Whether in the laboratory or the clinic, physi-cians and those who work with them are flat-footed positivists. Theirs is a pragmatic stance, rather like our stubborn adherence to the conceit (no matter how well we understand Copernican cosmology) that the sun rises and sets. Emphasizing practice over theory, medicine militates against deconstructionism in any strong sense: diseases may be socially constructed insofar as dif-ferent cultures, different sensibilities will recognize, describe, and even experience them differently. But in medicine the existence of disease itself is not in doubt; nor is the reality of the body, however socially prescribed our perception and experience of embodiment. Bodies may be understood culturally and historically, but for medicine they are palpably real. Pain, although it cannot be objectively measured, is reified for those who experience it and for those who have a duty to respond. This view of the world, especially in the United States, has weaned philosophers in medical ethics and literary scholars in medicine from their theoretical preoccupations and focused them instead on medical education and clinical practice, insulating them from late twentieth-century ideas *(50)*.

The effects of this medical positivism have been different in philosophy and in literature, especially in the relation of each to its mainstream discipline of origin. Reflecting the deep divisions in US departments of philosophy, philosophers in medical ethics for the most part have been allied with Anglo-American analytic philosophy and dismissive of developments in continental philosophy. Allusion to the later Wittgenstein is as wild as they get. By contrast, literature and medicine, isolated at first from English departments and from continental philosophy's influence on literary theory, now has a mainstream reputation as a faintly trendy, promising area for the study of culture. Its scholars are, nevertheless, not deconstructionists—nor have they ever been. In this they more closely resemble colleagues in medical ethics than those in literature departments. As a consequence, the discipline of literature and medicine has been far less open to intellectual currents in mainstream literary studies than its close cousin, literature and science.

The pragmatic effect of medicine on philosophy and literature has affected their focus as well. Demands for answers in the real world of clinical problems and public policy—like the effect on physicians of similar demands in the care of patients—has kept most medical ethicists focused on topical questions. The commodification of health care and the undiminished acceleration of medical technology—the human genome project, transgenic xenotransplantation, cloning—have meant that philosophers in medicine, though moderately radical when they depart from mainstream philosophy to take up real-world questions, still hold as essential and unquestioned the analytic tools they learned as graduate students. Scholars in literature and medicine, by contrast, have used the luxury of their perceived irrelevance to hot ethical topics to explore the ideas of the late-twentieth century. The tasks assigned to philosophers in medical ethics thus have limited their openness to interdisciplinarity and to interesting ethical questions raised by literature and literary studies.

## Literary Studies and the Critique of Medical Ethics

Although, as Mark Kuczewski has argued, there is at least a *de facto* consensus among philosophers about narrative in medical ethics *(51)*, the literary–theoretical perspective on the status of knowledge in medical ethics is not yet well understood. Such epistemological questions are the persistent concern of literary studies. Indeed, because good things to read were capably supplied by teachers from other disciplines well before literary scholars joined medical faculties, this critical perspective may ultimately be the most important contribution of literary studies to the common medical–ethical endeavor. Literary theory, almost alone, has offered a critique of medical ethics *(52)*.

At first, this critique was implicit: literature addresses ethical problems, but differently. Fiction, poetry, drama, and autobiographical essays provide a context for moral dilemmas, complicate debate about the variable force of medical–ethical principles, and broaden the scope of public policy questions. These are the all but inevitable consequences of reading literature and not of any particular literary theory. Texts concerned with illness and doctoring compel discussion of the nature of medicine, the character of the physician, the definition of dis-

ease, the existence and explanation of evil in the world.
Chekhov's "Ward No. 6" is not only about the slippery definition
of sanity and the proper treatment of the insane; it also asks its
readers whether it is better to accept stoically the world as it is, or
to venture small and surely futile acts in the face of indifference
and neglect. Likewise, Williams' "The Use of Force," which
illustrates the conflict of ends and means created by a sick and
uncooperative child, is also a personal ethical morbidity-and-
mortality conference. The physician-narrator's account of the
event raises questions about the selfhood of a physician, the place
of anger and confession in a service profession, and the buried
sexuality of the patient–physician encounter.

In the 1990s the critique of medical ethics became more
overt, spreading through the medical humanities. As commonly
practiced, medical ethics was described as narrow and
deductivist: not just bloodless and hyper-rational in tone, but mis-
taken about the power and authenticity of its objective stance.
I'm as big a fan of objectivity as the next American. I want refer-
ees with clear sight, judges who attend to the law and shun bribes
of the subtlest kind. But where is objectivity to be found? Where
grounded? A joke about major-league umpires illustrates the difficulty. Three of baseball's finest are sitting in a bar. The first
says proudly, "I calls 'em like I sees 'em." The second takes a
swig of his beer and says defiantly, "I calls 'em like they *are*."
The third sets down his glass and says slowly, "They ain't nothing till I calls 'em."

This recognition of inescapable subjectivity is not relativism: the umpire doesn't call 'em arbitrarily or in a vacuum. The
game has rules; two other umpires are on the field; players and
managers are quick to protest bad calls—to say nothing of the
fans. The game is broadcast on radio and television, and pitches
and calls are commented upon; instant replay provides retrospec-
tion from several camera angles; tomorrow's papers will carry an
account of the game. Beyond the moment, there is a guild of

umpires; the game has a history; there is a history of just such calls; this umpire has his own record.

Here are the core issues with which literary studies grapples: problems of representation, their implications for the status of knowledge, the assertion of truth or knowability, and the contribution of the knower (his or her experience, psychology, or culture) to the known. These are philosophical questions too, but in the United States in the second half of the twentieth century, they did not much interest philosophers drawn to medical ethics. During that time, historians, anthropologists, even economists and psychologists, struggled with the grounds and reliability of their knowledge. Relinquishing their claims to be sciences (or at the very least proto-sciences) with varying degrees of regret and relief, these disciplines worked out satisfactory solutions to the perceived threat of relativism *(53)*. But questions of subjectivity and the ethics of representation have been almost entirely neglected in medical ethics. In this it resembles medicine, but without medicine's modernist, practical excuse.

The most direct challenge to medical ethics has come from narratology and the rhetorical study of ethics case construction. In 1994, Tod Chambers published the first of a series of papers that applied the concepts of literary theory to the central narrative genre of medical ethics, the case. Heretofore, medical ethicists (and literary scholars too) regarded the ethics case like a little laboratory, a think-problem in which a difficulty is analyzed so as to determine its solution. But every case, Chambers points out, occludes details that are as significant as those it highlights. The case is presented from a narrative stance in a distinctive voice, and the narrator, direct or implied, inevitably makes assumptions about the world and the narratability of events. "To ignore the narrative characteristics that the medical ethics case shares with fiction is to confuse representation with the thing it represents—to mistake the story for the reality—and thus to miss the theory in the case" *(54)*. There is no pure, objective presentation of a case, and though there may be a cultural or national or professional consensus on the values engaged by a case and principles that apply to it, further examination, reinterpretation, and retelling are never foreclosed. The medical ethicist's case, far from being a piece of the world isolated for the testing of assumptions and hypotheses, Chambers shows, has been constructed from the very materials it purports to test.

Philosophers who mistake this critique of the medical ethics case as a criticism of themselves fail to grasp the point. Subjectivity is the inescapable condition of human knowing; all our science and much of our intellectual life is an attempt to correct for it. They ask, "What do you recommend?" "How can we fix it?" Chambers commends a narratological competence for medical ethicists that resembles the narrative competence Charon advises for physicians and ethicists *(55)*. They as authors (and we as readers) need to be aware of the rhetorical constructedness of the case that is among their best tools *(56)*. Such awareness is little different from that required of historians and intelligent readers of history. Philosophers with profit might follow the lead of historiographers—and more recently casuists *(57)*—to ask why medical ethics needs narrative in its search for truth; what part case-narrative plays in relation to its other tools of inquiry and explanation; and how it is related to the principles of medical ethics and other forms of moral knowledge.

The genuine problem for both disciplines, indeed for intellectual life as a whole, is how to give proper weight to subjectivity. Hilde Lindemann Nelson has asked the important questions. How can an ethicist honor the personal without being arbitrary? If the particulars are important, what about the general? *(58)* Or, as Arras puts it, near his conclusion, "We all need to think much harder about how to acknowledge our individuality and situatedness without abandoning the possibility of social criticism" (p. 84) These are important questions in literary theory but absolutely vital ones in medicine, where a cardinal virtue is the equal treatment of all comers. Not that it is always practiced, but so central is this virtue that hospitals and third-party payers have learned to set up barriers to the examination rooms. Once there, a person, no matter who, becomes a patient and the physician's best efforts are called forth *(60)*. The relation of particulars to the general is troublesome beyond the question of access to care: How can physicians safely engage their emotions in their practice? How can such a practice be unbiased? Can decisions be made case by case without ultimate unfairness to some group of patients? These are interpretive questions that can be ignored, but they do not go away. They have a practical parallel in the clinical use of generalities in diagnosis. On the one hand, forty-year-old women very seldom have heart disease; on the other, this particular forty-year-old woman is complaining of chest pain. Because medicine is a practice and not, like medical ethics, primarily a discourse, it allows for some slippage between rules and actions. In clinical medicine, rules often conflict, as do principles. The rule about rules is that they not be arbitrary, but the act may fudge it if the situation warrants and circumstances allow. In a good hospital, the woman will have an EKG.

Surely the best solution is a reciprocal engagement of both particulars and the general, both the concrete details and the abstractions. Whether it is called a hermeneutic circle or the achievement of a reflective equilibrium, the point is both, not either-or *(61)*. This sort of practical rationality is central to clinical medicine and to ethical decisionmaking. Medicine properly practiced is ethics in action. Analysis is needed when problems arise. But narrative and interpretive skills are essential to recognize problems, to understand them so as to attempt a solution, and to know whether a solution has been reached. It is difficult to do this as an outsider to the culture—not because cultures have different principles (although they often do), but because the meaning of the principles is determined in the cultural world where patients (and physicians and ethicists) live.

## Interdisciplinary Prospects

From the point of view of literature and literary theory, a genuine interdisciplinarity in medical ethics is a desirable but distant prospect. Although the strengths of medical ethics are considerable, they are limited in this regard. Its chief strength is the marvelous clarity philosophical training brings to moral discourse—although in human affairs, such clarity is probably best seen as a temporary way station and not a readily achieved goal. Medical ethics is also interdisciplinary in principle, and a recent openness (of which this collection of essays is an example) raises hope that it may move from the misplaced scientism that seems to follow from its narrow view of rationality to a new, more contextual way of understanding moral problems in medicine. For the most part, however, real interdisciplinarity lies in the future.

Literature and literary studies are not a panacea for the narrowness of medical ethics. They bear with them various weaknesses of their own. They do not speak the language of philosophy or marshal arguments in the same way. Sometimes they seem not to argue at all. The influence of continental philosophy has reinforced this tendency, and as a consequence, literary studies are ignored—or, worse, invited and then dismissed *(62)*. But even if the two fields could engage one another, there is much about literary studies that is inimical to philosophy. Literature and literary studies are messy, complexified unto contrariness. Their texts, whether in print or elsewhere, seem interminably interpretable, and this instability can undermine rule-based answers and their justification. Worse, the recent interest in narrative in the medical humanities seems to encourage a sentimental, almost pious attitude toward patients and the patient's story—an attitude that short circuits critical attention to the need for diagnosis (and moral judgment), and even to good narrative practice itself. On the whole, literature seems a suspect means of introducing emotion and subjectivity into rational discourse.

Guilty as charged. But of course I see most of these flaws as strengths. They are not a replacement for philosophy as practiced in medical ethics but its necessary complement. For philosophical medical ethics also has its weaknesses, ones well matched to the strengths of literature and medicine. These weaknesses include a reluctance to see knowledge as inevitably situated and contextual, a rush to judgment on questions of policy and practice that neglects the opportunity to educate participants, a lack of interest in the relation of its theory to its practice, and a widely held assumption that emotion is irrational. All stem from the monocular privileging of logico-mathematical rationality, a move that renders the use of narrative and narrative rationality in ethics officially invisible. The result is the failure of an intellectual (and too often a practical) interdisciplinarity. These are weaknesses for which, in a genuine dialogue, literature and literary theory can offer some complementary strengths.

First, literary studies is already interdisciplinary, almost promiscuously so. So too is medical ethics, but while its practitioners tend not to acknowledge, literary studies revel in it. No text interprets itself, and every commentary on a text becomes available for interpretation in its turn. Therefore (like that other interpretive enterprise, the law), literary studies draws on history, philosophy, economics, psychology, sociology, anthropology, and religious studies—whatever comes to hand. In many instances, literary studies simply borrow back what was earlier lent. One of "blurred genres" Clifford Geertz recommended to social scientists was narrative, and in the last thirty years, anthropology, historiography, and legal studies especially, have been strongly influenced in their method and concept of rationality by narrative and narrative theory.

Second, knowledge in literature is richly detailed, contextual, and inescapably situated. Stories, poems, and drama are crammed with information we may not think we need. Mere atmosphere! But, as with the umpire's call, context guides our interpretation. The parents' shy passivity in William Carlos

Williams' story is a shred of justification for the narrator's use of force; would he have acted in the same way with a prosperous or assertive family? These days it is fairly well accepted that literature offers richer accounts of moral knowledge than are customarily found in sociology or medical ethics, but the corollary is less well understood. The moral knowledge provided by literature is never simple, always particular, and inevitably situated in time and place, and by the rhetoric of narration. No text lacks the subtly graded frames of author, implied author, and narrator, reader and implied reader (just to give the simplest version of this Chinese box) *(63)*. Even when, as is often the case, such subtleties are unfamiliar to the reader, they work to convey the central problem of knowing outside the sciences: who is telling us? How does she know? How is her perspective coloring her representation? The practical understanding of ethical questions from the point of view of literary studies is thus a matter of interpreting the accounts of the participants and working out with them, the best possible next chapter. Principles are guidelines or, better, as John Dewey described them, hypotheses to be tried in (and by) these circumstances *(64)*. The hit-and-run provision of an answer without discussion with the patient, the family, and those who have taken care of the patient, uniformly condemned by other approaches, is literally inconceivable in narrative ethics. Ethics consultation in hospitals, on this view, gives way to education, and, although it may facilitate conclusions, it avoids dispensing advice or rendering judgments.

Third, literature limits abstraction, generalization. Universalizability is the hallmark of a just decision, but conclusions reached through narrative are not always universalizable. As Herman Melville's *Billy Budd* persuades us, the presence of rich detail and the inescapable situatedness of all narration make it difficult to determine relevant circumstances. Like the legal process and casuistry generally, the novel keeps the conversation open—not only about the death penalty, but about the grounds on which decisions are made and the problem of universalizability itself.

Fourth, as an aspect of this openness, literature always bears the possibility of interrogating the uses to which it is put. Its centrality to contextual thinking in medical ethics and to ethics education in medicine is due not only to its rich presentation of ethical dilemmas but also to the questions it raises about how these dilemmas are identified and resolved. As a consequence, literary studies offer an example of reflexive inquiry. Reading, which models the practice of interpretation, is subject to theoretical scrutiny, while literary theory, because it is "experience-near," is subjected to the test of practice. Values may authorize the reinterpretation of a text; equally likely, a rereading—like a rewriting of history—may call into question or ignore the very conclusions or values that had seemed so obvious before. A comparable attention in medical ethics to assumptions about knowledge and representation, now peripheral, would be interestingly productive.

Fifth, literature offers its readers experience with a broader concept of rationality, an alternative to the monocular focus on hypothetico-deductive rationality prized in science, analytic philosophy, and, too often, medical ethics. We have just begun to understand narrative as an alternative rationality: much work remains to be done particularly in the ethnography of medicine, in philosophy and, especially, in neurobiology. But it is certain that the account of rationality derived from the analytic tradition now current in medical ethics is too narrow to be useful in considering the larger moral matters engaged by medicine. Not that ethicists do not use narrative: they do so frequently and effectively. But they do not acknowledge it; deductivism remains the "gold standard" of ethical rationality. Physicians may have the obverse experience. While they may find a well-argued piece on, say, the moral equivalence of withholding and withdrawing treatment entirely convincing, it can be all but useless in their practice. Something doesn't feel right; the argument does not take account of their experience.

Sixth, literature authorizes the rationality of emotion. Its representation of emotion renders it available for observation and analysis. Nor is emotion separable from moral reasoning. Nussbaum has argued that "cognitive activity...centrally involves emotional response. We discover what we think...partly by noticing how we feel"; Julia Connelly describes the use of poetry to extend this attention to the clinical encounter *(65)*. Certainly, literature presents both the patient and the physician as full human beings. Poetry, drama, fiction, and autobiography tell us what it is like to die or to break the news to a dead patient's family. They allow us to glimpse human beings suffering the full weight of life's misfortune, whereas others supply the medicine and advice that may alleviate it.

These complementary strengths enable a truly interdisciplinary medical ethics to offer an alternative to the misplaced scientism of principlism as usual. Nothing is wrong with a good answer to a problem. Consensus is often reached on an issue in medical ethics, and in that sense progress is made. But, contrary to the view of some medical ethicists, most issues are far from settled, and such settlements that have been reached are open to review and revision. The ongoing discussion of the fine points of informed consent or the distinction between withdrawing and withholding treatment should be a sign that such issues are situational. Every moral problem has a history and an immediate social context that includes much about the agents that we cannot know *(66)*. Principles, then, are most usefully regarded as hypotheses. Like diagnoses, they must be demonstrated anew each time, or at least survive the skepticism of those investigating the matter. This practice, akin to Dewey's pragmatic fallibilism, blurs the distinction between ethics as education and ethics as practice. The debates over ethics consultation—who is qualified and how, and how to measure a consultation's success or failure—might disappear as consultative conclusions written in the chart become less important than ethical discussion that

preceded them. The field would be genuinely open to good practitioners who, almost necessarily, would also be good teachers.

Medical ethics is an ongoing, dialogic, socially and historically conditioned discourse about practical decisions in our society. Ideally conducted, it is an inclusive, multivocal enterprise
open to all comers and to all languages of argument and description. As education and mail delivery are handed over by the rich
or well-funded to private suppliers, medical care comes close to
being a culture-wide (if not uniform) phenomenon. For all its
horrid inequities, it may be the last, most nearly democratic institution in our time. Discussion of decisions in the medical arena
may be our best chance of sustaining a society-wide conversation
about issues that matter. We close off medical ethics from knowledge gained at the movies or from poems like Rafael Campo's
"Ten Patients and Another" at our peril.

How human beings know what is right and, before that, how
we recognize events and situations as morally problematic are
matters that lie deeper than their logical representation. Although
hypothetico-deductive reasoning is comfortingly systematic and
undoubtedly useful in dealing with moral quandaries, the recognition and understanding of those quandaries, like our knowledge
of culture and its values generally, is part of a more discursive,
practical, and narrative rationality. A good physician, like other
reliable moral agents, grasps not just the solution to an ethical
dilemma but the action appropriate to morally significant situations. This larger, contextual moral interest, so integral to the
practice of medicine, can be split off from medical ethics. Some
might argue that this split has so often occurred (or been called
for) in recent years; that it is an implicit goal of bioethicists. But
such a split ultimately divides medical ethics from medicine as a
moral practice and deprives medical ethics of the insights of literary theory that are its most interesting and powerful critique. Literature then would be (as philosophers have continued to
conceive of it) the untheorized handmaiden of medical ethics:
merely good illustrations of moral dilemmas that interrupt pro-

fessional life, rather than, as it truly is, the source and method of moral knowledge in our culture.

## Acknowledgment

A somewhat different version of this essay appeared in the *Hastings Center Report* 31 (2001), 36–43.

## Notes and References

[1] *See* Anne Hunsaker Hawkins. (1993) *Reconstructing Illness: Studies in Pathography*. Purdue University Press, West Lafayette, IN; and Arthur W. Frank. (1994) "Reclaiming an Orphan Genre: The First-Person Narrative of Illness". *Literature and Medicine* 13: 1–21.

[2] *See*, to name only a sample. Selzer's *Letters to a Young Doctor*, Stone's *In All This Rain* (LSU Press, 1981); Klass' *Other Women's Children*; Campo's *What the Body Told*; Stein's *The White Life,* Hall's *Without*; Moore's "People Like That Are the Only People Here," in *Birds of America*; and Hoffmann's *"Complaints of a Dutiful Daughter."* In these works, first-person narrators are not reliable signs of autobiography any more than declared autobiography guarantees unmediated fact: truth to experience is required of all.

[3] Lewis Thomas (1974) describes its "half-way technology" in *Lives of the Cell.* Bantam, New York, pp. 35–42. Michel Foucault ponders its centrality in *The Birth of the Clinic: An Archeology of Medical Perception*, (trans. A.M. Sheridan Smith) Vintage, New York, pp. 196–198.

[4] Anthony Moore. (1978) *The Missing Medical Text.* University of Melbourne, Melbourne.

[5] Tom L. Beauchamp and James F. Childress. (1994) *Principles of Biomedical Ethics,* 4th ed. Oxford University Press, New York.

[6] Joanne Trautmann. (1981) *Healing Arts in Dialogue: Medicine and Literature.* Southern Illinois University Press, Carbondale, IL.

[7] One philosopher who is the exception is Edward Pincoffs. (1986) *Quandaries and Virtues: Against Reductivism in Ethics.* University Press of Kansas, Lawrence. See especially "Quandary Eth-

ics," pp. 13–36. William May's (1983) makes a case against "quandary ethics" in *The Physician's Covenant: Images of the Healer in Medical Ethics.* Westminster, Philadelphia.

[8]Robert Coles. (1979) "Medical Ethics and Living a Life." *N. Engl. J. Med.* 301: 444–446.

[9]Anne Hudson Jones. (1987) "Literary Value: The Lesson of Medical Ethics." *Neohelicon* 14: 383–392. In addition, the theologian William May, contributed essays on current theater to the *Hastings Center Report*, for years the only reference to literature in its pages.

[10]In *After Virtue: A Study in Moral Theory* (1981) Alasdair MacIntyre describes moral life, not as decisions made at crucial choice points but as the whole narrative trajectory of an individual life: Notre Dame University Press, Notre Dame, IN, especially pp. 199–200.

[11]My first hint that there might be a chink in this formalist armor came in 1961 when Mark Schorer returned to a modern British novel class after a day spent in court on behalf of Grove Press, which had illegally published D. H. Lawrence's banned *Lady Chatterly's Lover* in the US. He described his testimony supporting the argument that art has no effect on morals. Then he shrugged, noting that if literature could not admit its ill effects, it would have a hard time claiming its good ones. Subsequent historians of literary theory have noted how convenient the formalist position was during the McCarthy (and late segregationist) era when, by contrast, the teaching and writing of historians came under frequent scrutiny.

[12]Horace. *The Ars Poetica of Horace.* (1960) (ed., Augustus S. Wilkins) Macmillan, New York.

[13]Kathryn Montgomery Hunter. (1986) "The Physician as Textual Critic," in *The Connecticut Scholar: Occasional Papers of the Connecticut Humanities Council* 8: pp. 27–37; Rita Charon. (1989) "Doctor-Patient/Reader-Writer: Learning to Find the Text," in *Soundings* 72: pp. 137–152.

[14]See *Literature and Medicine* 5 (1986) and 7 (1988).

[15]Suzanne Poirier and Daniel J. Brauner. (1990) "The Voices of the Medical Record," in *Theoretical Medicine* pp. 29–39; and Poirier et al., "Charting the Chart: An Exercise in Interpretation(s)," in *Literature and Medicine* 11: 1–22; Kathryn Montgomery Hunter.

(1991) *Doctors' Stories: The Narrative Structure of Medical Knowledge*. Princeton University Press, Princeton., William J. Donnelly. (1988) "Righting the Medical Record: Transforming Chronicle into Story." *JAMA* 260: 823–825; and (with Daniel J. Brauner) (1992) "Why SOAP Is Bad for the Medical Record". *Archives of Internal Medicine* 152: 481–484. All of us participated in the Chicago Narrative Reading Group, which, with Ann Folwell Stanford, twice presented workshops on the medical case history—"Re-Weeding the Record"—at meetings of the Society of Health and Human Values.

[16]See, for example, Stanley Cavell. (1969) "The Avoidance of Love," in *Must We Mean What We Say?* Cambridge University Press, Cambridge; Iris Murdoch, (1971) The Sovereignty of Good. Schocken, New York; Bernard Williams. (1981) *Moral Luck: Philosophical Papers 1973–1980.* Cambridge University Press, New York.

[17]Jerome Bruner. (1986) *Actual Minds, Possible Worlds* (Harvard University Press, see especially Chapter 2, "Two Modes of Thought," pp. 11-43.

[18]Roger C. Schank., (1990) T*ell Me a Story: A New Look at Real and Artificial Memory.* Scribners, New York. See also Roger C. Schank and Robert P. Abelson. (1981) *Scripts, Plans, Goals and Understanding: An Inquiry into Human Knowledge.* Erlbaum, Hillsdale, NJ.

[19]Martha C. Nussbaum. (1986) *The Fragility of Goodness: Luck and Ethics in Greek Tragedy and Philosophy.* Cambridge University Press, Cambridge. Her study of the moral themes in the fiction of Henry James amplifies the argument that literature is essential to addressing questions of the moral life: *Love's Knowledge: Essays on Philosophy and Literature.* (1990) Oxford University Press, New York.

[20]Wayne C. Booth. (1988) *The Company We Keep: An Ethics of Fiction*. University of California Press, Berkeley,.

[21]Lawrence Buell describes it as a "groundswell," "even if it has not—at least yet—become the paradigm-defining concept that textuality was for the 1970s and historicism [was] for the 1980s," in "Introduction: In Pursuit of Ethics" [Special Topic], *Publication of the Modern Language Association* (1999), 114: 7–19.

[22]See for example J. Hillis Miller. (1987) *The Ethics of Reading: Kant, de Man, Eliot, Trollope, James, and Benjamin.* Columbia University Press, New York.

[23]David P. Haney. (1999) "Aesthetics and Ethics in Gadamer, Levinas, and Romanticism: Problems of Phronesis and Techne." *PMLA* 114:32–45; see also Adam Zachary Newton. (1995) *Narrative Ethics.* Harvard University Press, Cambridge, a study of modern literature that draws on Bakhtin, Levinas, and Cavell. Neither work has anything at all to do with biology or medicine.

[24]Jerome Bruner. (1990) *Acts of Meaning.* Harvard University Press, Cambridge.

[25]I stumbled onto the blurred distinction when I wrote a book about medicine that was read as a book about medical ethics, and slowly learned to appreciate the value of the confusion. I'm indebted to Mary Mahowald for insisting I had something to say to her medical ethics course at the University of Chicago; to Mark Waymack for an invitation to speak at a Society for Health and Human Values meeting in April of 1995; and to Tod Chambers, who relentlessly poked fun at my replying to his brainstorming on ethics cases with ruminations about medical ones.

[26]*See*, for example, Anne Hudson Jones. "From Principles to Reflective Practice or Narrative Ethics?" a reply to Ronald A. Carson's "Medical Ethics as Reflective Practice," both in (1997) *Philosophy of Medicine and Bioethics: A Twenty-Year Retrospective and Critical Appraisal.* (R.A. Carson and Chester R. Burns, eds.) Kluwer Academic Publishers, Dordrecht, pp. 181–195.

[27]Rita Charon, Joanne Trautmann Banks, Julia E. Connelly, Anne Hunsaker Hawkins, Kathryn Montgomery Hunter, Anne Hudson Jones, Martha Montello, and Suzanne Poirier. (1995) "Literature and Medicine: Contributions to Clinical Practice." *Annals of Internal Medicine* 122: 599–606. This article, like the *Academic Medicine* article cited next, was a part of the project Charon undertook to remedy the paucity of theoretical justification in literature and medicine.

[28]Kathryn Montgomery Hunter, Rita Charon, and John Coulehan. (1995) The Study of Literature in Medical Education." *Academic Medicine,* 70: 787–794. This article was one of eight devoted to the contributions of philosophy, literature, history, religious stud-

ies, ethics, art, and law to medicine education in a special section edited by Rita Charon and Peter Williams.

[29]See Albert R. Jonsen and Stephen Toulmin. (1988) *The Abuse of Casuistry: A History of Moral Reasoning.* University of California Press, Berkeley; Richard M. Zaner. (1994) "Experience and Moral Life: A Phenomenological Approach to Bioethics," in *A Matter of Principles? Ferment in US Bioethics.* (Edward R. DuBose, Ron Hamel, Laurence J. O'Connell, eds.) Trinity Press International, Valley Forge, PA, pp. 211–239; Drew Leder. "Toward a Hermeneutical Bioethics," in *A Matter of Principles? Ferment in US Bioethics.* (DuBose et al., eds.) pp. 240–259. Joseph J. Fins, Matthew D. Bacchetta, and Franklin G. Miller. (1997) "Clinical Pragmatism: A Method of Moral Problem Solving." *Kennedy Institute of Ethics Journal* 7: 129–145.

[30]Rita Charon. (1986) "To Render the Lives of Patients," in *Literature and Medicine* 5: 58–74. Her example has been well followed. Douglas R. Reifler has instituted writings across the medical curriculum at Northwestern; *see* (1996) "I don't actually mind the bone saw." *Literature and Medicine* 15: 183–199. See also John L. Coulehan, Peter C. Williams, D. Landis, and Curt Nasser. (1995) "The First Patient: Reflections and Stories about the Anatomy Cadaver." *Teaching and Learning in Medicine* 7: 61–66; and Martin Kohn and Joseph Zarconi. (1995) "'I am not stupid, I was a school teacher'": A Narrative Approach to Teaching Clinical Medical Ethics." *Reflections: Narratives of Professional Helping* 1: 52–58.

[31]Howard Brody. (1987) *Stories of Sickness.* Yale, New, Haven; Arthur Kleinman. (1988) *The Illness Narratives: Suffering, Healing and the Human Condition.* Basic Books, New York.

[32]Steven H. Miles and Kathryn Montgomery Hunter. (1990) "Case Stories." *Second Opinion* 15: 60–69; the series continued three times a year through volume 19 (1993).

[33]Rita Charon. "Narrative Contributions to Medical Ethics: Recognition, Formulation, Interpretation, and Validation in the Practice of the Ethicist," in *A Matter of Principles? Ferment in US Bioethics* (DuBose et al., eds.) pp. 260–283.

[34]Anne Hudson Jones. (1997) "Literature and Medicine: Narrative Ethics." *Lancet* 349: 1243–1246.

[35]Thomas H. Murray. (1997) "What Do We Mean by 'Narrative Ethics?'" in *Stories and Their Limits: Narrative Approaches to Bioethics.* (Hilde Lindemann Nelson, ed.) Routledge, New York, pp. 3–17.

[36]K. Danner Clouser. (1996) "Philosophy, Literature, and Ethics: Let the Engagement Begin." *Journal of Medicine and Philosophy.* 21:324–328.

[37]Tom Tomlinson. (1988) Perplexed about Narrative Ethics, in *Stories and Their Limits*, (H. L. Nelson, ed.) pp. 123–133. *See also*, James S. Terry and Peter C. Williams, Literature and Bioethics: The Tension in Goals and Styles. *Literature and Medicine* 7:1–21.

[38]Joanne Trautmann Banks. (1990) Literature as a Clinical Capacity: Commentary on "the Quasimodo Complex." *Journal of Clinical Ethics* 1: 2273–231; Anne Hunsaker Hawkins, *Literature, Medical Ethics*; and Epiphantic Knowledge.' *Journal of Clinical Ethics* 5: 283–290.

[39]David B. Morris. (1995) *Illness and Meaning in the Postmodern Age.* University of California, Berkeley; Arthur W. Frank. *The Wounded Storyteller.* University of Chicago Press, Chicago.

[40]John Arras. Nice Story, But So What? Narrative and Justification in Ethics, in *Stories and Their Limits.* (H. L. Nelson, ed.) pp. 65–88. Subsequent page numbers in the text refer to the essay.

[41]David Burrell and Stanley Hauerwas. (1989) From System to Story: An Alternative Pattern for Rationality, in, *Why Narrative? Readings in Narrative Theology.* (Stanley Hauerwas and L.G. Jones, eds.) Eerdmans, Grand Rapids, MI, pp. 158–190. For MacIntyre, *see* note 10 above.

[42]Arras is defending principles against an attack that, at least in narrative ethics, has not occurred. No one wants to throw out principles: not Hauerwas or MacIntyre, as Arras demonstrates; not even Arthur Frank, who is "doing ethics" not ethical theory when he suspends judgment to accord every patient an equal voice. Medical–ethical principles need to be understood in their social and historical context. As the very American rallying cries of relatively powerless observers, principles remind powerful agents of their values and their duty. Essential to a new field, they are still valuable. They should not be thrown out, but they are not foundational. They provide guidance, not answers, which seems to me

the way it should be. The goal is not universal answers, after all, but lively discourse—professional, interdisciplinary, civic—and a reflective consideration of moral problems should not be limited to philosophers who speak a special language. Their analytical skills will still be useful, even vital, as all reductive skills are when clarity of focus is needed. Those times occur often, but they're not the whole of moral life; not even the most important parts, however interesting they may be to those who possess those skills. If a philosopher regards this restriction of ethics to the exercise of philosophic skills and language as "preserving" medical ethics, that is understandable. If a literary scholar regards such a claim to "preservation" as so much turf-protection, she hopes to be found understandable too. If medical ethics is to be interdisciplinary, there will be no handmaidens.

[43]This term for the unity of bioethics and medical practice is from Charon's introduction to Rita Charon, Howard Brody et al., Literature and Ethical Medicine: Five Cases from Common Practice, (1996) *Journal of Medicine and Philosophy* 21: 243–247. There she draws a distinction between medicine and medical ethics, but especially in a collection of narratives of medical practice, it is a distinction without meaningful difference to physicians.

[44]Stephen Toulmin. (1982)How Medicine Saved the Life of Ethics. *Perspectives in Biology and Medicine* 25: 736–750.

[45]They are not the first, of course: besides Sacks and Selzer, there are David Hilfiker's examinations of conscience on such scrupulously avoided topics as mistakes and poverty in *Healing the Wounds* (1985) Pantheon, New York; and A. Gonzales-Crusi's brooding meditations on human embodiment. But recent essays are distinguished by their number and accessibility: although they are non-fiction, they are aimed at a non-medical audience, published in books and in the New Yorker rather than in the *New England Journal of Medicine*.

[46]*See* especially Charles LeBaron's (1982) *Gentle Vengeance*. Penguin, New York; and Perri Klass' (1987) *A Not Entirely Benign Procedure*. Putnam,New York.

[47]Clifford Geertz. (1983) Blurred Genres: The Refiguration of Social Thought, in *Local Knowledge: Further Essays in Interpretive*

*Anthropology*. Basic Books, New York, pp. 19–35.

[48]Clouser, pp. 323–324.

[49]Charles Taylor. (1989) *The Sources of the Self: The Making of Modern Identity*. Cambridge University Press, Cambridge,.

[50]Medical ethics in Europe is more open to other philosophical approaches and almost inevitably more aware of national variation in disease labels and therapy. *See* Henk ten Have. Principlism: A Western European Appraisal, in *A Matter of Principles*. (DuBose et al., eds.) pp. 101–120.

[51]Mark Kuczewski. Bioethics' Consensus on Method: Who Could Ask for Anything More? in *Stories and Their Limits*. (H. L. Nelson, ed.) pp. 134–149. The consensus is seldom acknowledged. Like prose, that wonderful thing Moliere's Bourgeois Gentilhomme discovers he has been speaking all along, narrative in ethics has been invisible, "natural," and until recently, unanalyzed.

[52]David Rothman's *Strangers at the Bedside* remains the only history of bioethics; there is no ethnography. I have argued that other disciplines—especially history and historiography, cultural anthropology, art theory—could have offered such a challenge and speculated about why thus far they have not; *see* Kathryn Montgomery Hunter. (1996) Literature, Narrative, and the Clinical Exercise of Practical Reason. *Journal of Medicine and Philosophy*. 21: 303–320.

[53]In addition to Geertz's "Blurred Genres," and Bruner's "Two Modes of Thought," cited above (notes 24 and 27), *see* Hayden White. (1986) The Value of Narrativity in the Representation of Reality, in *The Content of the Form: Narrative Discourse and Historical Representation*. Johns Hopkins University Press, Baltimore, pp. 130–151; and Donald Spence. (1982) Narrative Truth, Historical Truth: Meaning and Interpretation in *Psychoanalysis*. Norton,New York.

[54]Tod Chambers, (1996) From the Ethicist's Point of View: The Literary Nature of Ethical Inquiry. *Hastiness Center Report* 26:25–32; *see also* The Bioethicist as Author: The Medical Ethics Case as Rhetorical Device (1994) *Literature and Medicine* 13: 60–78; and *The Fiction of Bioethics* (1999) Routledge, New York.

[55]Rita Charon, "Narrative Contributions," pp. 275–276.

[56]Tod Chambers. What to Expect from an Ethics Case (and What It

Expects from You), in *Stories and Their Limits*. (H.L. Nelson, ed.) pp. 171–184.

[57]*See* John D. Arras. (1994) Principles and Particularities: The Roles of Cases in Bioethics. *Indiana Law Journal* 69: 992 ff. Howard Brody has addressed these questions for "ethical medicine," *see The Healer's Power* (1992) Yale, New Haven.

[58]Hilde Lindemann Nelson. Introduction: How to Do Things with Stories, in *Stories and Their Limits*. (H. L. Nelson, ed.) pp. vii–xx.

[58]There is some evidence that the ethos of nursing is different; *see* " The Patient's Story," forthcoming *Medical Anthropology Quarterly*.

[60]This is the conclusion to the comparable problem in clinical reasoning reached by Trisha Greenhalgh in her *British Medical Journal* essay, "Narrative Based Medicine in an Evidence Based World" (January 1999), collected in *Narrative Based Medicine: Dialogue and Discourse in Clinical Practice*. (Trisha Greenhalgh and Brian Hurwitz, eds.) BMJ Books, London, pp. 247–265.

[61]Literature and Medical Ethics, (1996) special issue of *J. Med. Philos.* (K. Danner Clouser and Anne Hunsaker Hawkins, eds.) 21: 3.

[62]Carl Elliott. (Jul–Aug, 1992) Where Ethics Comes From and What to Do About It. *Hastings Center Report* pp. 28–35; a later version appears as Chapter 8: (1999) A General Antitheory of Bioethics, in *A Philosophical Disease: Bioethics, Culture, and Identity*. Routledge, New York, pp. 141–164.

[63]Wayne C. Booth. (1983).*The Rhetoric of Fiction, 2nd ed.* University of Chicago Press, Chicago.

[64]John Dewey. (1988), The Nature of Principles, in *Human Nature and Conduct [1922] in The Middle Works, 1899–1924*, vol. 14. Southern Illinois Press, Carbondale pp. 164–170. Howard Becker adopts Dewey's idea in his phronesiology of social science, *Tricks of the Trade: How to Think about Your Research While You're Doing It*. (1998) University of Chicago Press, Chicago.

[65]Nussbaum *The Fragility of Goodness*, pp. 15–16; Julie Connelly. (1994) Being in the Present Moment: Developing the Capacity for Mindfulness in Medicine. *Academic Medicine* 74: 420–424.

[66]*See* Tod Chambers, (1993) "Voices;" Steven Miles, "Ms. Lubell's Complaint;" and Kathryn Montgomery Hunter, (1993) "The Whole Story," *Second Opinion* 19(2): 81–103.

# 7

# History and Bioethics

*M. L. Tina Stevens*

## Introduction

Historian Charles Rosenberg cautions bioethicists that they cannot be self-aware "without an understanding of the history of medicine in the past century." If they ignore history, he warns, bioethics will be unable to situate "the moral dilemmas it chooses to elucidate." Bioethics will become a "self-absorbed technology, mirroring and eventually legitimating that self-absorbed and all-consuming technology it seeks to order and understand"*(1)*. His advice speaks to the vital role that history should generally play in the bioethical enterprise. But how and if history, as an academic discipline, did or could assist the flourishing of bioethics—a central concern of this anthology—is a question about which history, as an academic discipline, could be rather indifferent. For even if bioethics were to fade away, its three-to-four decade existence is still a fascinating topic, historically speaking. How did it come to be, Why did it last only as long as it did, and Why did it decline? are just a few of the larger questions that could inspire historical speculation and research for generations. For chroniclers, accounting for how and why bioethics

From: *The Nature and Prospect of Bioethics: Interdisciplinary Perspectives*
Edited by: F. G. Miller, J. C. Fletcher, and J. M. Humber
© Humana Press Inc., Totowa, NJ

declined would be as irresistible a project as explaining its genesis and growth. Moreover, historical accounts by academic historians (as opposed to reflections by the participants themselves) cannot really be said to have assisted in the initial flourishing of bioethics since history is, by definition, an *ex post facto* investigation. Bioethical histories could, of course, influence the continued unfolding of bioethics as historical considerations of the field proliferate and are read by bioethicists. But how or if the historian's craft does or ever will help promote bioethical flourishing must be left to the assessments of future bioethical enthusiasts. For now, it is safest to recount the historiography of bioethics, to reflect on what has been of historical interest so far, what remains of interest, and how we might approach thinking about and researching it historically.

## Historiography

The historiography of bioethics reveals the birthmarks of a fledgling field: participant histories and professional historian accounts focused intently on origins *(2)*. Rosenberg reminds us that histories generated by participants in a developing field can serve self-celebratory ends which can mystify as much as analyze *(3)*. While this does, indeed, characterize some participant accounts, it does not wholly describe the not-completely-celebratory critique offered by sociologist Renee Fox, one of the early members of the world's first bioethics institute, the Hastings Center *(4)*. In what is chiefly a sociological consideration, Fox offers a brief historical explanation for the emergence of bioethics: it grew out of the 1960s as a response to scientific and medical technological developments. "Bioethics...surfaced in American society in the late 1960s," she relates, "a period of acute social and cultural ferment. From its outset, the value and belief questions with which it had been preoccupied have run parallel to those with which the society had been grappling more broadly." Later, she explains more specifically that "growing professional and public concern about moral aspects of experimentation with human subjects, particularly in the sphere of medical research...played the major triggering role in the genesis of bioethics"*(5)*. Implicating its technological determinism, she explains that bioethics had come to focus on problems associated with "a particular group of advances in biology and medicine," with, she adds, "strikingly little acknowledgment of the improvement in identifying, controlling, and treating disease that these advances represent" *(6)*. Explaining what she means by "technological determinism" she clarifies that:

> Much of the bioethical literature is based on the assumption that the value questions that have arisen in the field of biomedicine have been "caused"or "created" by medical, scientific and technological advances. Partly because of its biomedical and technological determinism, bioethical analysis does not usually take note of the fact that some of the same cultural questions that have crystallized around biological and medical developments have also been central to many non-medical issues. *(7)*

From simple institutional origins in the late sixties consisting chiefly of the activities of pioneering centers like the Hastings Center, the Kennedy Institute, and the Society of Health and Human Values, bioethics quickly "pervaded the public domain," resonating in courtrooms, national commissions, and the media. Although the field has changed somewhat since its inception, the original "ethos of bioethics," according to Fox, was wedded to "individual rights, autonomy, self-determination, and their legal expression in the jurisprudential notion of privacy," as well as to the values of truth-telling, distributive justice, cost containment, and the principle of beneficence. The weight bioethics gives to the value of individualism limits the field. Fox criticizes that bioethics "has relegated more socially oriented values and ethical questions to a secondary status." This tendency, along with its clear technological determinism, bends bioethics "away from involvement in social problems" *(8)*. Bioethics, Fox concludes, is conservative in important ways.

> [T]he way that bioethics has defined and focused on the value complex of individualism, the degree to which it has played down a social perspective on personal and communal moral life, its parsimonious acceptance of a cost-containing framework of health care analysis, and the extent to which its rationality and methodology have distanced it away from the phenomenological reality of medical ethical situations have converged to form the gestalt that is congruent with other fundamentals of a conservative outlook. *(9)*

Fox charges bioethics with paring down the complexity of social problems to fit within a utilitarian, positivist, and reductionist framework. Historian David Rothman does not share Fox's negative assessment of bioethics' focus on individual rights. In fact, in *Strangers at the Bedside*, Rothman urges that it was precisely bioethics' strong commitment to individual rights that gave it such broad social appeal *(10)*. Bioethics' commitment to patients'' rights and autonomy helped it to fashion a "new alliance among outsiders to medicine" that challenged the discretionary authority of the medical profession. Fox and Rothman also seem to disagree over the significance of the class location of bioethics. Fox views the professional, scholarly and academic orientation of bioethicists to be at odds with the more "grassroots" understanding of issues of those outside the "upper middle class professional and guild enclaves" *(11)*. Their own professional and upper class status, combined with their lack of concern for social context when considering bioethical issues, constitutes bioethicists as a conservative group. For Rothman, however, bioethics was not so much blinded by its class as it was freed by its commitment to individual rights, to cross class lines altogether. Bioethics, argues Rothman, was, "at least as responsive, and perhaps even more so, to the concerns of the haves than the have-nots.

Not everyone is poor or a member of a minority group or disadvantaged socially and economically; but everyone potentially, if not already, is a patient" *(12)*.

But, like Fox, Rothman agrees that bioethics is a child of the sixties and a product of that decade''s larger concern with civil rights. Bioethics shared in the hallmark struggle of the era: that of the individual against "constituted authority":

> The fit between the movement and the times was perfect. Just when courts were defining an expanded right to privacy, the bioethicists were emphasizing the principle of autonomy, and the two meshed neatly. ...[J]ust when movements on behalf of a variety of minorities were advancing their claims, the bioethicists were defending another group that appeared powerless—patients. All these advocates were siding with the individual against the constituted authority; in their powerlessness patients seemed at one with women, inmates, homosexuals, tenants in public housing, welfare recipients, and students, who were all attempting to limit the discretionary authority of professionals. *(13)*

Rothman eschews the idea that bioethics grew chiefly out of worries over biotechnologies. It was not a crude technological determinism that gave bioethics its social purchase power, he exhorts, but its attitude of challenge to medical authority. Ethical abuses in the area of experimentation with human subjects that came to light in the in the 1960s forced this challenge—an attack waged by bioethicists and other outsiders against the bastion of authority that had been the "doctors'' preserve." Rothman''s narrative, however, does not address the fact that important calls for ethical oversight came from within the medical and scientific research communities themselves, a fact which weakens the characterization that bioethics' *raison d'etre* was its outsiders' challenge *(14)*.

In *The Birth of Bioethics*, Albert Jonsen, a bioethicist and an early participant in the nascent field, seems to agree with Rothman's view that American postwar liberalism and the Civil

Rights movement constituted the breeding ground of bioethics. He shares, in his last chapter, how he himself was active in the Civil Rights movement as were a number of other early bioethicists *(15)*. Curiously, however, despite the invocation of the cultural importance of the Civil Rights Movement at the end of his book, the main narrative of events on which Jonsen focuses is not intimately infused with a strong sense of the influence of this 1960s cultural hallmark. In fact, the historical framework that Jonsen offers at the beginning of his work designates the more general period between 1947 and 1987 as "the era during which bioethics emerged as a distinct discipline and discourse." And although these years include the period that cradled the Civil Rights movement, by starting with the immediate postwar period, Jonsen implicates other historical events more directly.

Jonsen explains that he chose 1947 because in that year, "...the Nuremberg Tribunal convicted twenty-three physicians of war crimes committed under the guise of medical experiments, and it promulgated the Nuremberg Code. ...[I]t initiated an examination by professional persons in science, medicine, and law of one of modern medicine's central features: scientific research." The centerpiece of Jonsen's text chronicles five "topics that became the focus of bioethical analysis: research with human subjects, genetics, care of terminally ill persons, organ transplantation, and artificial reproduction." Although he chooses 1947 as marking the beginning of the bioethics era, the narratives of these five topics, start "long before 1947, in the conviction that the bioethical shape of these modern problems must be seen within the evolution of thought about their analogues in the past" *(16)*.

But Jonsen himself acknowledges that American medical research in general, felt largely unchallenged by the concerns of Nuremberg, believing that a code based on Nazi behavior could impart little wisdom not already possessed by civilized physicians. Additionally, he recognizes that the efforts of the few medical men who did speak out publicly regarding ethical challenges facing American medical researchers resulted in no institutional changes. Moreover, he agrees that it was not until the late 1960s that philosophers or theologians turned seriously to the ethical problems of medical research *(18)*. Yet his narrative of events does not explicitly puzzle out the historical question as to why there was a roughly twenty-year gap between the 1947 Nuremberg Code and the actual emergence of bioethics in the late 1960s. Just over twenty years after Nuremberg, ethicists felt the need to come to terms with what they viewed as ethical dilemmas being generated by advances in biomedicine. The efforts to determine why early bioethicists believed that there was a revolution and to determine, historically, the sources of the revolution are different from the effort of simply tracing back the geneologies of topics that would, eventually, become the staples of bioethics as a discipline (human subjects research, genetics, care of teminally ill persons, organ transplantation, or artificial reproduction). In other words, Jonsen was correct when he characterized his effort as finding past *analogs* to bioethical issues. But the historical task, I would argue, is one of locating the *homologous* roots of bioethics, of explaining why ethicists could make believable claims that a new expertise—a new profession—was necessary in order to adequately scrutinize technologies and biomedical procedures that then seemed so troubling.

In *Bioethics in America: Origins and Cultural Politics*, I try to sort through this question by taking seriously the early statements of high profile bioethicists warning society that it needed to concern itself urgently with unprecedented ethical questions then being generated by what they called "the biological revolution"*(18)*.

This approach is consistent with Fox's assessment of bioethics' technological determinism. My research suggests that there are both long-term and more recent cultural and historical sources of bioethics. Taking a long gaze back through American history, it is possible to see that for many generations, members of the educated elite found certain aspects of technological development disturbing. Although critical of technological development, they operated within the larger social class that fundamentally supported and generated scientific, medical, and technological research and development. Ultimately, they and their current intellectual descendents function to ease development of what are considered exotic biotechnologies by diffusing more virulent dissent against it. The most recent cultural and intellectual forerunners of bioethicists emerged during an era of acute modernization between 1880 and 1920. The disquietude some intellectuals felt over the modern technological nature of the dominant society led them to turn to alternative cultural havens like medievalism, orientalism, and "primitivism." Historian T. J. Jackson Lears explains how these "antimodernists" were at once critical of material development but also eager for it. This "half-commitment," he suggests, worked ultimately to dissipate critiques of the culture:

> Half-committed to modernization, antimodernists unwittingly allowed modern culture to absorb and defuse their dissent. Unable to transcend bourgeois values, they often ended by revitalizing them. Ambivalent critiques became agenda for bourgeois self-reformation: antimodern craft ideologues became advocates of basement workbench regeneration for tired corporate executives; antimodern militarists became apologists for modern imperialism... Antimodern thinkers played a key (albeit often unknowing) role in revitalizing the cultural hegemony of their class during a protracted period of crisis. *(19)*

Given the nature of the claims made by bioethicists, their socioeconomic location, and, arguably, how they function to marginalize more radical critiques of biotechnological development, one is drawn to the conclusion that bioethicists constitute the late-twentieth century's version of this phenomenon. In the context of the 1960s, bioethicists not only diffused their own cultural critique but also functioned to diffuse the more radical variety of dissent by popular intellectuals like Herbert Marcuse, Jacques Ellul, and Theodore Roszak.

In postwar America, the most high-profile expression of this ambivalent posture toward exotic scientific development was the responsible science movement that surfaced in the wake of atomic detonation. Strong evidence suggests that the members of the scientific and medical communities who requested assistance from ethicists and theologians at what can be seen as the beginning of bioethics in the 1950s, had been influenced by the atomic scientists' calls for social responsibility. This implicates an earlier birth date for bioethics than what is suggested by more dominant views; it also underscores that "lay" advice into medico-scientific dilemmas was initially invited by the scientific and medical communities; it was not a case of ethicists storming the battlements.

In addition to suggesting an earlier genesis of bioethics than previous accounts have done, and emphasizing the role of requests for "lay" advice coming from within biomedical communities, *Bioethics in America* also offers an alternative view of the cultural function of bioethics. Where some earlier assessments view bioethics as a challenge to biomedical authority or at least as offering potentially threatening outsider oversight of it, *Bioethics in America* concludes that bioethics plays an important role in buttressing biomedical authority by midwifing the ultimate social acceptance of exotic biotechnological development.

Charles Rosenberg also highlights the "ambiguous role" that bioethics has played when he comments on the way bioethics legitimizes authority by questioning it:

> ...from the historian's perspective, [bioethics] has played a complex and in some ways ambiguous role. Bioethics not only questioned authority, it has in the past quarter-century helped constitute and legitimate it. As a condition of its acceptance, bioethics has taken up residence in the belly of the medical whale; although thinking of itself as still autonomous, the bioethical enterprise has developed a complex and symbiotic relationship with this host organism. *(20)*

The conclusions I reached after analyzing archives at the Hastings Center and examining bioethical responses to the 1968 redefinition of death (one of the first developments for which fledgling bioethicists were asked for their *imprimatur*) and its aftermath in the 1975 *Quinlan* case, indicate that bioethics operates to quell a society suffering from what Alvin Toffler diagnosed at the dawn of the 1970s as the disease of change—"future shock." Toffler defines this disease as "the shattering stress and disorientation that we induce in individuals by subjecting them to too much change in too short a time" *(21)*. Bioethics as a social institution (not every bioethicist practicing individually) has functioned to ease society into the acceptance of exotic biotechnologies that, on first impression, engender alarm *(22)*.

Although *Bioethics in America* casts a broader historical net than previous accounts of the rise of bioethics, it has been criticized for not casting even further—for ignoring the supposed influences of such developments as feminism or the ecology movement in giving rise to bioethics *(23)*. But this type of criticism stems from the fallacy of *post hoc ergo propter hoc*, that is, believing that what occurred before an event must have caused it. Certainly, important intellectual and social developments surfaced just before, or in tandem with, bioethics, but not all of them were equally influential in creating the concept of the "biological revolution," a characterization that gave bioethicists their *raison d'etre*, buttressed their efforts toward institutionalization, and bestowed great purchase toward social acceptance. By examining what "pre-bioethicists" themselves were reading in the 1950s and early 1960s, and paying attention to what they considered to be influential in generating a biological revolution, *Bioethics in America*, follows a paper trail to a number of trigger concerns troubling postwar scientists, doctors and, eventually, ethicists. In turn, many of those they influenced later would claim the appellation "bioethicist" and undertake their vigil over the biological revolution. This constitutes a historical methodology that is very different from simply asserting that parallel developments surely must have given birth to bioethics.

As bioethics has matured into a social institution, it has made an effort to distance itself from charges of adhering to a crude technological determinism; it is perhaps for this reason that invocation of a biological revolution has subsided somewhat. Indeed, the biological revolution as an explicitly stated concept plays no consciously analyzed role in Jonsen's rendition of bioethics' birth, even though the force of his narrative is driven, in large part, by the debuts of various biomedical technologies *(24)*.

## Directions for Future Inquiry

The topical areas for archival research and historical methodology and inquiry into bioethics are vast, whatever the overarching thematic approach. Rosenberg suggests that there are three "social spaces" that bioethics inhabits: the academic, the institutional (hospital and research settings), and the media *(25)*. Such a structural understanding can help indicate areas for historical investigation. Additionally, there is likely to be continuing interest in the origins and historical social function of bioethics.

Eventually, as historical considerations of bioethics proliferate, efforts will be taken to periodize it in order to understand its growth and development. Already we have seen that there is disagreement over what constitutes its origins. There will be different ways of marking more subtle trends and developments as well. In her 1989 essay, for example, Fox suggested that bioethics had already been through three different phases between the late-1960s and the mid-1980s, whereas Jonsen designates the entire period from 1947 to 1987 more broadly as bioethics' emergent era *(26)*. There are endless ways to characterize change and to argue for what constitutes a new era. I have noticed that bioethicists seem to have moved away (although not entirely) from their initial explicit alarm over needing to deal with ethical challenges presented by the biological revolution—perhaps because now that they are firmly established, they need no longer justify their existence, perhaps because society no longer requires convincing. But whatever the reason, the lower decibel level of this concern suggests that bioethics is no longer in its infancy. Also suggesting important changes in bioethical development is Hastings Center founder Daniel Callahan's recent twofold observation: that bioethics became "secularized" early on when it broadened it's theological roots to include lawyers, philosophers, etc., and changed again when it moved away from an early position which cherished the possibilities for maintaining "neutral ethics" to feeling that it is "now acceptable in a way it was not twenty years ago" to commit to causes, to "come down on this or that side of an issue" *(27)*.

Beyond a concern with periodization, a few themes stand out for continuing inquiry. A central channel of the most recent historical and sociological critiques of bioethics echoes Fox's early evaluation when they emphasize how bioethical commitment to "principalism," that is, to the use of abstractions like autonomy, beneficence, nonmaleficence and justice when offering ethical assessments, serves to decontextualize and depoliticize situations that are inherently social and political *(28)*. This causes bioethicists to overlook not only the political nature of ethical situations they may be called on to assess but also to miss the political dimensions of their own historical development. Additionally, bioethics tends to ignore the inherently political nature of biotechnological development itself.

In their essay "Why Bioethics Needs Sociology," Raymond DeVries and Peter Conrad discuss bioethical "blind spots." They emphasize that "[l]ack of sensitivity to the structure of medical care systems prevents American bioethicists from seeing the way they protect the status quo." They urge (in agreement with Daniel Chambliss) that ethicists and ethics committees serve the interests of medical organizations by deflecting attention away from structural deficiencies in health care, redefining them as limited ethical problems." They advise that wisdom from the field of the sociology of professions and from sociological ethnographies more generally may help redress this imbalance *(29)*.

Moving through similar corridors of inquiry, historians are likely to continue emphasizing the importance of social, political, and cultural context of bioethics. Rosenberg, for example, recently stressed the importance of remembering the political, cultural, and economic contexts in which bioethical issues inhere:

> ...we cannot remove or isolate value assumptions from the institutional, the technical, and the conceptual in medicine... Medicine is negotiated and inevitably political and...the political is cultural. ...Questions that can be framed as matters of justice and autonomy are at once questions of control and economic gain. Perceptions of right and wrong, of appropriate standards constitute de facto political realities. *(30)*

As historical analysis continues to "contextualize" bioethics, it will, in essence, be making inquiries into how and why bioethicists came into existence and thrived by studying biotechnologies which they construed as autonomously developed social products, that is, entities produced without political influence or political ramifications, and by believing themselves to be autonomous—free from political influence.

When bioethicists speak of "autonomy" in the course of their practice, reference is typically being made to one of their guiding principles, namely, having respect for the personal liberty of an individual *(31)*. But the concept of "autonomy" figures into at least two other aspects of bioethics that likely will be of continuing interest to "bioethics watchers." The first aspect concerns whether biotechnology can ever really be considered autonomous, that is, developed without political influence or implication, as so much of bioethical practice and theory tacitly assumes. There is evidence that some early bioethicists had been influenced by intellectuals concerned with the philosophy of technology, the likes of Lewis Mumford, Jacques Ellul, or Herbert Marcuse *(32)*. In their attachment to the tools of moral philosophy, however, bioethicists let slip away the early influence of this other philosophical area. The inquiry begun by these social critics in the nineteen fifties and sixties and continued more recently by those few who, like Langdon Winner, explore some of the "ways in which conditions of power, authority, freedom, and social justice are deeply embedded in technical structures," is largely ignored by bioethicists today *(33)*.

A second aspect of autonomy worthy of continued historical inquiry (intimately related to the first) is whether the ethical deliberations of bioethicists can ever be unaffected by—autonomous from—the influence of those who fund them or accept their expertise. "For a time in the seventies," Daniel Callahan tells us, "many in the field held that bioethics should be neutral, providing dispassionate analysis not partisan judgment. That view didn't last long, a victim of the soon-perceived impossibility of doing "'neutral ethics'" *(34)*. Callahan was speaking of the challenges facing bioethicists in trying to maintain an even-handed, non-ideological posture when deliberating on bioethical issues. This problem is compounded, however, when bioethicists receive funds from corporate sources, a difficulty recently given public airing in a high profile article featured in the *New York Times*. "Some bioethicists accept corporate donations for their university programs," the article explained, "and others work as paid consultants for biotechnology companies leading colleagues to charge that they are being used as public relations tools and damaging the field's credibility" *(35)*. In *Bioethics in America*, an account is offered of the difficulties encountered by the early Hastings Center in attempting to nurture impartiality by freeing itself from financial ties and points out issues surrounding funding obtained from any source, whether corporate, university, governmental, or private as targets for continuing socio-historical inquiry.

## Conclusion

Typically, decades pass before an event is considered ripe for historical consideration. As such, history, as an academic discipline, is a late-comer to the interdisciplinary field of bioethics. Now, some thirty years since the emergence of the first bioethics organizations, histories are beginning the work of accounting for the institutionalization of what is a decided social phenomenon. When and why, did the field develop? What is its cultural function? How has the field changed over the course of its brief existence? Already, there are competing answers to these questions. How bioethicists themselves answer these questions will be part of continuing historical interest in this fascinating area. And if bioethicists do not choose to find history as essential to the development of their field, history, at least, will continue to find bioethics a compelling social development worthy of historical pursuit.

## Acknowledgment

The author teaches in the History Department at San Francisco State University. She is grateful to the following for their editorial assistance: Rosann Greenspan, Stephen Shmanske, Peggy Stevens, and Gloria Jeanne Stevens.

## Notes and References

[1]Charles E. Rosenberg. (1999, Fall) Meanings, Policies, and Medicine: On the Bioethics Enterprise and History, in *Daedalus, vol. 128, no. 4. See* pp. 27–46. Rosenberg's warning includes the importance of heeding lessons from sociological ethnographies and politics as well.

[2]Of course, what constitutes origins can be debated. (*See* below.)

[3]Rosenberg, p. 38.

[4]Of Fox's work, *see* especially, "The Sociology of Bioethics," in Renee Fox, *The Sociology of Medicine: A Participant Observer's View.* (1989) Prentice Hall, Inc., Englewood Cliffs, NJ . For her assessment of how bioethics has developed, *see*, "The Entry of U.S. Bioethics into the 1990s," in *A Matter of Principles: Ferment in US Bioethics.* (1994) Trinity Press International, Valley Forge, Pennsylvania; and "Is Medical Education Asking Too Much of Bioethics?", in *Daedalus.* (1999, Fall) Vol. 128, no. 4, pp. 1–25.

*See also*, "Advanced Medical Technology: Social and Ethical Implications," (1976) *Annual Review of Sociology 2*; and, Fox with Judith P. Swazey, (1988) "Medical Morality is Not Bioethics: Medical Ethics in China and the United States," in *Essays in Medical Sociology*. (ed. Renee C. Fox) New Brunswick, NJ.

[5]Fox (1989), pp. 224,225.

[6]Fox (1989), p. 225.

[7]Fox (1989), pp. 230,231.

[8]Fox (1989), pp. 229–234.

[9]Fox (1989), p. 233.

[10]Rothman engaged Fox's 1976 and 1988 articles while I drew my summary of Fox's critique from her 1989 article. As such, my comparison of their views here is based on my assessment of their works independently, as well as constituting a report of their engagement of each others ideas. (*See* note 5 for a list of articles by Renee Fox.)

[11]Fox (1989), p. 233 citing (1986) personal letter from Ruel Tyson, Professor in the Department of Religious Studies, at the University of North Carolina at Chapel Hill.

[12]David J. Rothman. (1991) *Strangers at the Bedside: A History of How Law and Bioethics Transformed Medical Decision Making.* Basic Books, p. 246.

[13]Rothman, p. 245.

[14]*See* M. L. Tina Stevens (2000) *Bioethics in America: Origins and Cultural Politics.* Johns Hopkins University Press, Baltimore, pp. 38–41.

[15]Jonsen, p. 387.

[16]Albert R. Jonsen. *The Birth of Bioethics* Oxford University Press, New York, p. xii.

[17]Jonsen, Chapter. 5, especially pp. 137,140,148.

[18]M. L. Tina Stevens (2000) *Bioethics in America: Origins and Cultural Politics* Johns Hopkins University Press, Baltimore, MD.

[19]T. J. Jackson Lears. *No Place of Grace: Antimodernism and the Transformation of American Culture:1880–1920.* (1981) Pantheon Books, New York,, pp. 57,58.

[20]Rosenberg, pp. 37–38.

[21]Alvin Toffler. (1971) *Future Shock.* Bantam Books, New York, p. 2.

[22]Historically, related social phenomena may be the Great Awakenings

of the 18th and 19th centuries, if those developments are understood as psycho-social responses to challenges presented by rapid
social change occurring within the more religious context of early
American society. Also, perhaps not unrelated to the idea of bioethics functioning as a social coping strategy, is Joseph
Campbell's observation that contemporary society is moving too
fast to develop or adhere to a mythology capable of making existence comprehensible. *See* Apostrophe S Productions, et al. (1988)
*Joseph Campbell and the Power of Myth, with Bill Moyers.*

[23]Terri Peterson. (2001, June 6) Book Review of "Bioethics in
America." *JAMA* Vol. 285, no. 21.

[24]*See* Stevens, pp. 30,31 for more general discussion.

[25]Rosenberg, p. 40.

[26]"The Sociology of Bioethics," in *Renee Fox, The Sociology of Medicine: A Participant Observer's View.* (1989) Prentice Hall,
Englewood Cliffs, NJ, p. 226.

[27]Daniel Callahan. (1997, Apr) "Bioethics and the Culture Wars," in
*The Nation,* p. 24.

[28]See for example, Paul Root Wolpe. (1998) "The Triumph of
Autonomy in American Bioethics: A Sociological View," in *Bio-
ethics and Society: Constructing the Ethical Enterprise.* Prentice
Hall, Inc, NJ, pp. 38–59.

[29]Raymond DeVries and Peter Conrad. (1998) "Why Bioethics Needs
Sociology," in *Bioethics and Society: Constructing the Ethical
Enterprise.* Prentice Hall, Inc, NJ, , p. 236. Other essays in this
volume emphasize the importance of ethnographic studies. *See
also,* Arthur Kleinman. (1999, Fall) "Moral Experience and Ethical Reflection: Can Ethnography Reconcile Them? A Quandary
for 'The New Bioethics,' in *Daedalus,* Vol. 128, no. 4, pp. 69–
97. Charles Bosk suggests, however, that "ethnographies...may
very well cut against the objectives of bioethicists. There may be
a built-in incompatibility between bioethical and sociological
inquiry, and heightening this tension rather than attempting to
deny it may very well be a useful contribution of the social scientists to bioethics." *See* his, "Professional Ethicists Available:
Logical, Secular, Friendly," in *Daedalus,* (1999 Fall), Vol. 128,
no. 4, pp. 47–68 at p. 65. *See also,* Arthur Kleinman, "Moral
Experience and Ethical Reflection: Can Ethnography Reconcile

Them? A Quandary for 'The New Bioethics,'" in Daedalus, (1999, Fall), Vol. 128, no. 4, pp. 69–97. Another relevant volume is Barry Hoffmaster, ed., (2001, Fall) *Bioethics in Social Context*, pp. 69–97.

[30]Rosenberg, p. 35.

[31]Jonsen, pp. 334–338.

[32]For example, *see*, Stevens, p. 52; Jonsen, p. 386.

[33]Langdon Winner. (1986) *The Whale and the Reactor: A Search for Limits in an Age of High Technology.* The University of Chicago Press, Chicago,. *See also,* Andrew Feenberg. (1999) *Questioning Technology.* Routledge Press, London and New York.

[34]Callahan. (1997) *Bioethics and the Culture Wars*, p. 24.

[35]Sheryl Gay Stolberg (2001) "Bioethicists Find Themselves The Ones Being Scrutinized," in *The New York Times*, August 2, p. 1.

# Index

liver, 81
organs, 54
Trilling, Lionel, 146

**U**

Uncertainty
  medicine, 72
Universalizability, 166
"Use of Force," 143

**V**

Virtues, 120

**W**

"Ward No. 6," 143, 160
Weakest sense of theory, 24
*Whole New Life*, 141
*Whose Life Is It Anyway*, 141
"Why Bioethics Needs
  Sociology," 190
Wikler, Dan, 31
Williams, Bernard, 148
Williams, William Carlos, 143
*Wounded Storyteller*, 153
Writers, 143
Wrongful disability, 22

# DATE DUE

| OCT 7 '03 | | | |
|---|---|---|---|
| OCT 1 2 '05 | | | |
| FEB 8 '06 | | | |
| | | | |
| | | | |
| | | | |
| | | | |
| | | | |
| | | | |
| | | | |
| | | | |
| | | | |
| | | | |
| | | | |
| | | | |